INTELLIGENT TRANSPORTATION SYSTEMS

Edited by **Ahmed Abdel-Rahim**

Intelligent Transportation Systems
Edited by Ahmed Abdel-Rahim

Published by InTech

Janeza Trdine 9, 51000 Rijeka, Croatia

Publishing Process Manager Molly Kaliman

Technical Editor Teodora Smiljanic

Cover Designer InTech Design Team

Additional hard copies can be obtained from orders@intechopen.com

Intelligent Transportation Systems, Edited by Ahmed Abdel-Rahim
p. cm.
ISBN 978-953-51-0347-9

For EU product safety concerns:
IN TECH d.o.o., Prolaz Marije Krucifikse Kozulić 3, 51000 Rijeka, Croatia,
info@intechopen.com or visit our website at intechopen.com.

Meet the editor

Dr Abdel-Rahim is an Associate Professor at the Civil Engineering Department, University of Idaho. He has authored and co-authored over 70 publications in academic journals, book chapters, conference proceedings, and research reports. Dr. Abdel-Rahim's expertise includes traffic control and operations, Intelligent Transportation Systems, microscopic simulation modeling. Hardware-in-the-loop simulation, network modeling, transportation network security and survivability, real-time control application, vehicle emission and fuel consumption modeling, highway and traffic safety, and engineering education.

Contents

Preface

With the rapid implementation of Intelligent Transportation System (ITS) applications throughout the world, the surface transportation system has become more complex and dependent on an extensive grid of roadways, computing devices, and wireless and wired communication networks. Authors from several countries have contributed chapters that focus on different components of ITS and their applications. Topics covered in the book include: ITS sensing devices, wirelesses communication networks to support ITS applications, ITS technologies for reduced vehicle emissions and environmental impacts, microscopic modeling of ITS systems, active ITS traffic management practices, and ITS applications in developing countries. The open exchange of ITS-related scientific results and ideas will hopefully lead to improved understanding ITS systems and their design and operations and to a greater awareness of ITS technologies and their potentials and applications.

Dr. Ahmed Abdel-Rahim, Ph.D., PE,
Civil Engineering Department,
University of Idaho,
USA

1

Applying Vehicular Networks for Reduced Vehicle Fuel Consumption and CO_2 Emissions

Maazen Alsabaan[1,3], Kshirasagar Naik[1], Tarek Khalifa[1] and Amiya Nayak[2]
[1]*University of Waterloo,*
[2]*University of Ottawa*
[3]*King Saud University*
[1,2]*Canada*
[3]*Saudi Arabia*

1. Introduction

These days the detrimental effects of air pollutants and concerns about global warming are being increasingly reported by the media. In many countries, fuel prices have been rising considerably. In western Canada, for instance, the gasoline price almost doubled from about 53 cents/liter in 1998 to 109 cents/liter in 2010 (Wiebe, 2011). In terms of the air pollution problem, greenhouse gas (GHG) emissions from vehicles are considered to be one of the main contributing sources. Carbon dioxide (CO_2) is the largest component of GHG emissions. For example, in Japan in 2008, the amount of CO_2 emissions from vehicles (200 million ton) is about 17 percent of the entire CO_2 emissions from Japan (1200 million ton) (Tsugawa & Kato, 2010). The Kyoto Protocol aims to stabilize the GHG concentrations in the atmosphere at a level that would prevent dangerous alterations to the regional and global climates (OECD/IEA, 2009). As a result, it is important to develop and implement effective strategies to reduce fuel expenditure and prevent further increases in CO_2 emissions from vehicles.

A significant amount of fuel consumption and emissions can be attributed to drivers getting lost or not taking a very direct route to their destination, high acceleration, stop-and-go conditions, congestion, high speeds, and outdated vehicles. Some of these cases can be alleviated by implementing Intelligent Transportation Systems (ITS).

ITS is an integration of software, hardware, traffic engineering concepts, and communication technology that can be applied to transportation systems to improve their efficiency and safety (Chowdhury & Sadek, 2003). In ITS technology, navigation is a fundamental system that helps drivers select the most suitable path. In (Barth et al., 2007), a navigation tool has been designed especially for minimizing fuel consumption and vehicle emissions. A number of scheduling methods have been proposed to alleviate congestion (Kuriyama et al., 2007) as vehicles passing on an uncongested route often consume less fuel than the ones on a congested route (Barth et al., 2007).

Various forms of wireless communications technologies have been proposed for ITS. Vehicular networks are a promising research area in ITS applications (Moustafa & Zhang, 2009), as drivers can be informed about many kinds of events and conditions that can impact travel.

To exchange and distribute messages, broadcast and geocast routing protocols have been proposed for ITS applications (Broustis & Faloutsos, 2008; Sichitiu & Kihl, 2008) to evaluate network performance (e.g., message delays and packet delivery ratio), instead of evaluating the impact of the protocols on the vehicular system (e.g., fuel consumption, emissions, and travel time).

This chapter studies the impact of using a geocast protocol in vehicular networks on the vehicle fuel consumption and CO_2 emissions. Designing new communication protocols that are suitable in applications, such as reducing vehicle fuel consumption and emissions, is out of this chapter's scope. The purpose of this chapter is to:

- Motivate researchers working in the field of communication to design economical and environmentally friendly geocast (EEFG) protocols that focus on minimizing vehicle fuel consumption and emissions;
- Demonstrate the ability to integrate fuel consumption and emission models with vehicular networks;
- Illustrate how vehicular networks can be used to reduce fuel consumption and CO_2 emission in a highway and a city environment.

This research brings together three key areas which will be covered in Sections 2, 3, and 4. These areas are: (1) geocast protocols in vehicular networks; (2) vehicle fuel consumption and emission models; and (3) traffic flow models. Section 5 will introduce two scenarios where applying vehicular networks can reduce significant amounts of vehicle fuel consumption and CO_2 emissions.

2. Geocast protocols in vehicular networks

Geocast protocols provide the capability to transmit a packet to all nodes within a geographic region. The geocast region is defined based on the applications. For instance, a message to alert drivers about congestion on a highway may be useful to vehicles approaching an upcoming exit prior to the obstruction, yet unnecessary to vehicles already in the congested area. As shown in Figure 1, the network architectures for geocast in vehicular networks can be Inter-Vehicle Communication (IVC), infrastructure-based vehicle communication, and Hybrid Vehicle Communication (HVC). IVC is a direct radio communication between vehicles without control centers. Thus, vehicles need to be equipped with network devices that are based on a radio technology, which is able to organize the access to channels in a decentralized manner (e.g., IEEE 802.11 and IEEE 802.11p). In addition, multi-hop routing protocols are required, in order to forward the message to the destination that is out of the sender's transmission range. In infrastructure-based vehicle communication, fixed gateways are used for communication such as access points in a Wireless Local Area Network (WLAN). This network architecture could provide different application types and large coverage. However, the infrastructure cost has to be taken into account. HVC is an integration of IVC with infrastructure-based communications.

The existing geocast protocols are classified based on the forwarding types, which are either simple flooding, efficient flooding, or forwarding without flooding (Maihöfer, 2004). In this chapter, geocast protocols are classified based on performance metrics. An important goal of vehicular networks is to disseminate messages with low latency and high reliability. Therefore, most existing geocast protocols for vehicular networks aim to minimize message

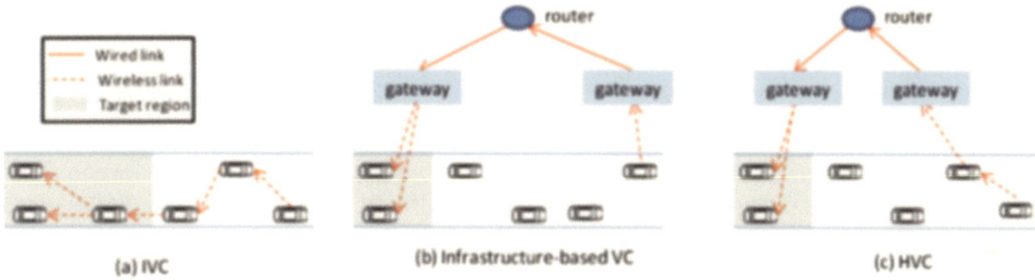

Fig. 1. Possible network architectures for geocast in vehicular networks.

latency, or to increase dissemination reliability. In this chapter, we want to draw the attention of researchers working in the field of communication to design geocast protocols that aim to reduce vehicle emissions.

2.1 Geocast protocols aim to minimize message latency

Message latency can be defined as the delay of message delivery. A higher number of wireless hops causes an increase in message latency. Greedy forwarding can be used to reduce the number of hops used to transmit a packet from a sender to a destination. In this approach, a packet is forwarded by a node to a neighbor located closer to the destination (Karp & Kung, 2000). Contention period strategy can potentially minimize message latency. In reference (Briesemeister et al, 2000), when a node receives a packet, it waits for a period of time before rebroadcast. This waiting time depends on the distance between the node and the sender; as such, the waiting time is shorter for a more distant receiver. The node will rebroadcast the packet if the waiting time expires and the node did not receive the same packet from another node. Otherwise, the packet will be discarded.

2.2 Geocast protocols aim to increase the dissemination reliability

One of the main problems associated with geocast routing protocols is that these protocols do not guarantee reliability, which means not all nodes inside a geographic area can be reached. Simple flooding forwarding can achieve a high delivery success ratio because it has high transmission redundancy since a node broadcasts a received packet to all neighbors. However, the delivery ratio will be worse with increased network size. Also, frequent broadcast in simple flooding causes message overhead and collisions. To limit the inefficiency of the simple flooding approach, directed flooding approaches have been proposed by

1. Defining a forwarding zone;
2. Applying a controlled packet retransmission scheme within the dissemination area.

Location Based Multicast (LBM) protocols are based on flooding by defining a forwarding zone. In reference (Ko & Vaidya, 2000), two LBM protocols have been proposed. The first protocol defines the forwarding zone as the smallest rectangular shape that includes the sender and destination region. The second one is a distance-based forwarding zone. It defines the forwarding zone by the coordinates of sender, destination region, and distance of a node to the center of the destination region. An intermediate node broadcasts a received packet only if it is inside the forwarding zone. Emergency Message Dissemination for Vehicular environment (EMDV) protocol requires the forwarding zone to be shorter than the communication range and to lie in the direction of dissemination (Moreno, 2007). The forwarding range is adjusted according to the probability of reception of a single hop broadcast message. In this case, high reception probability near the boundary of the range can be achieved.

A retransmission counter (RC) is proposed as a packet retransmission scheme (Moreno, 2007). When nodes receive a packet, they cache it, increment the RC and start a timer. RC=0 means the node did not receive the packet correctly. The packet will be rebroadcast if the time is expired. Moreover, the packet will be discarded if the RC reaches a threshold.

For small networks, temporary caching can potentially increase the reliability (Maihofer & Eberhardt, 2004). The caching of geounicast packets is used to prevent the loss of packets in case of forwarding failures. Another type of caching is for geobroadcast which is used to keep information inside a geographical area alive for a certain of time.

2.3 Geocast protocols aim to reduce vehicle fuel consumption and emissions

To the best of our knowledge, all existing protocols focus on improving the network-centric performance measures (e.g., message delay, packet delivery ratio, etc.) instead of focusing on improving the performance metrics that are meaningful to both the scientific community and the general public (e.g., fuel consumption, emissions, etc.). The key performance metrics of this chapter are vehicle fuel consumption and CO_2 emissions. These metrics can be called economical and environmentally friendly (EEF) metrics.

Improving the network metrics will improve the EEF metrics. However, the existing protocols are not EEF because their delivery approach and provided information are not designed to assist vehicles in reducing uneconomical and environmentally unfriendly (UEF) actions. These actions include

- Acceleration;
- High speed;
- Congestion;
- Drivers getting lost or not taking a very direct route to their destination;
- Stop-and-go conditions;
- Idling cars on the road;
- Choosing a path according to a navigation system that later becomes congested and inefficient after committing to that path.

3. Fuel consumption and emission models

A number of research efforts have attempted to develop vehicle fuel consumption and emission models. Due to their simplicity, *macroscopic* fuel consumption and emission models have been proposed (CARB, 2007; EPA, 2002). Those models compute fuel consumption and emissions based on average link speeds. Therefore, they do not consider transient changes in a vehicle's speed and acceleration levels. To overcome this limitation, *microscopic* fuel consumption and emission models have been proposed (Ahn & Rakha, 2007; Barth et al., 2000), where a vehicle fuel consumption and emissions can be predicted second-by-second. An evaluation study has been applied on a macroscopic model called MOBILE6 and two microscopic models: the Comprehensive Modal Emissions Model (CMEM) and the Virginia

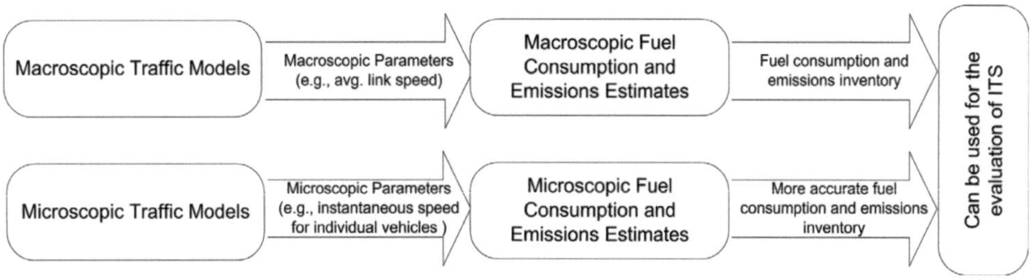

Fig. 2. Summary of the link between traffic flow and fuel consumption and emission models.

Tech Microscopic model (VT-Micro) (Ahn & Rakha, 2007). It has been demonstrated that the VT-Micro and CMEM models produce more reliable fuel consumption and emissions estimates than the MOBILE6 (EPA, 2002). Figure 2 shows the link between transportation models and fuel consumption and emissions estimates.

Microscopic models are well suited for ITS applications since these models are concerned with computing fuel consumption and emission by tracking individual vehicles instantaneously. The following subsections briefly describe the two widely used microscopic models.

3.1 CMEM model

The development of the CMEM began in 1996 by researchers at the University of California, Riverside. The term "comprehensive" is utilized to reflect the ability of the model to predict fuel consumption and emissions for a wide variety of vehicles under various conditions. The CMEM model was developed as a power-demand model. It estimates about 30 vehicle/technology categories from the smallest Light-Duty Vehicles (LDVs) to class 8 Heavy-Duty Trucks (HDTs) (Barth et al., 2000). The required inputs for CMEM include vehicle operational variables (e.g., second-by-second speed and acceleration) and model-calibrated parameters (e.g., cold-start coefficients and engine-out emission indices). The cold-start coefficients measure the emissions that are produced when vehicles start operation, while engine-out emission indices are the amount of engine-out emissions in grams per one gram of fuel consumed (Barth et al., 2000; UK, 2008). The CMEM model was developed using vehicle fuel consumption and emission testing data collected from over 300 vehicles on three driving cycles, following the Federal Test Procedure (FTP), US06, and the Model Emission Cycle (MEC). Both second-by-second engine-out and tailpipe emissions were measured.

3.2 VT-Micro model

The VT-Micro model was developed using vehicle fuel consumption and emission testing data obtained from an experiment study by the Oak Ridge National Laboratory (ORNL) and the Environmental Protection Agency (EPA). These data include fuel consumption and emission rate measurements as a function of the vehicle's instantaneous speed and acceleration levels. Therefore, the input variables of this model are the vehicle's instantaneous speed and acceleration. The model was developed as a regression model from experimentation with numerous polynomial combinations of speed and acceleration levels as shown in the following equation.

$$\ln(MOE_e) = \begin{cases} \sum_{i=0}^{3}\sum_{j=0}^{3}(L_{i,j}^{e} \times s^i \times a^j), & \text{for } a \geqslant 0 \\ \sum_{i=0}^{3}\sum_{j=0}^{3}(M_{i,j}^{e} \times s^i \times a^j), & \text{for } a < 0 \end{cases} \quad (1)$$

where

$\ln(y)$: Natural logarithm function of y, where y is a real number;

s: Instantaneous vehicle speed (km/h);

a: Instantaneous vehicle acceleration (km/h/s);

MOE_e: Instantaneous fuel consumption or emission rate (L/s or mg/s);

e: An index denoting fuel consumption or emission type, such as CO_2, HC, and NO_x emissions. e is not an exponential function;

$M^e_{i,j}$: Model regression coefficient for MOE_e at speed power i and acceleration power j for negative accelerations;

$L^e_{i,j}$: Model regression coefficient for MOE_e at speed power i and acceleration power j for positive accelerations.

As noticed from Equation 1, the model is separated for positive and negative accelerations because vehicles exert power in positive accelerations, while vehicles do not exert power in the negative accelerations. The VT-Micro model is inserted into a microscopic traffic simulator called "INTEGRATION" to compute vehicles' fuel consumption and emissions (Van, 2005a;b). This model has been used in this research due to its simplicity and high accuracy since it produces vehicle emissions and fuel consumption that are consistent with the ORNL data. The correlation coefficient between the ORNL data and the model predicted values ranges from 92% to 99% (Ahn et al., 2002).

3.2.1 Example of using the VT-Micro model

Sample model coefficients for estimating fuel consumption rates for a composite vehicle are introduced in Table 1. The composite vehicle was derived as an average across eight light-duty vehicles. The required input parameters of the model are:

- Instantaneous speed (km/h);
- Instantaneous acceleration (km/h/s);
- Model regression coefficient for positive and negative acceleration as given in Table 1.

Consider a vehicle started traveling. A microscopic traffic model has to be utilized in order to measure the vehicle instantaneous speed and acceleration. Simulation of Urban Mobility (SUMO) has been used in this regard. SUMO is a microscopic traffic simulation package developed by employees of the Institute of Transportation Systems at the German Aerospace Center (Krajzewicz et al., 2002).

VT-Micro model has a speed-acceleration boundary. For instance, at speed 50 km/h, the maximum acceleration that can be used in the model is around 2.2 m/s^2 (Ahn et al., 2002). In this example, the maximum vehicle speed, acceleration and deceleration are set to 50 km/h, 2 m/s^2 and -1.5 m/s^2, respectively. The second-by-second speed and acceleration are computed for the first 5 seconds of the vehicle's trip as shown in Table 2. It is noticed that all accelerations are positive. By applying the input parameters to Equation 1, the fuel consumption estimates should be as demonstrated in Table 2 and Figure 3. Clearly from the table, the increase or decrease of the fuel consumption is based on the speed and acceleration. Although fuel

Coefficients	s^0	s^1	s^2	s^3
Positive a				
a^0	-7.73452	0.02799	-0.0002228	1.09E-06
a^1	0.22946	0.0068	-0.00004402	4.80E-08
a^2	-0.00561	-0.00077221	7.90E-07	3.27E-08
a^3	9.77E-05	0.00000838	8.17E-07	-7.79E-09
Negative a				
a^0	-7.73452	0.02804	-0.00021988	1.08E-06
a^1	-0.01799	0.00772	-0.00005219	2.47E-07
a^2	-0.00427	0.00083744	-7.44E-06	4.87E-08
a^3	0.00018829	-0.00003387	2.77E-07	3.79E-10

Table 1. Sample VT-Micro model coefficients for estimating fuel consumption

Time (s)	1	2	3	4	5
Speed (km/h)	7.2	13.356	18.648	23.184	27.072
Acceleration (km/h/s)	7.2	6.156	5.292	4.536	3.888
Fuel Consumption (liter)	0.002338176	0.002502677	0.00260202	0.002611232	0.002555872

Table 2. Instantaneous speed, acceleration and fuel consumption

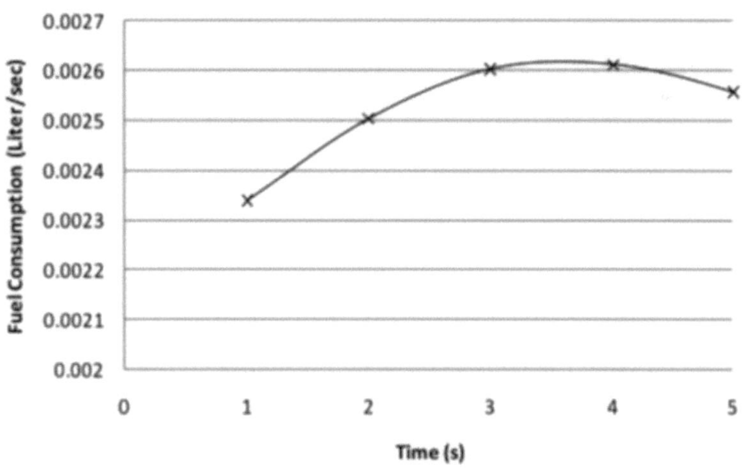

Fig. 3. Instantaneous fuel consumption.

consumption normally increases with increasing acceleration, it is not the largest amount at the time of the highest acceleration, which is 6.156 km/h/s at the 2nd second because of the speed effect on the fuel consumption. Likewise, at the highest speed, which is 27.072 km/h at the 5th second, the fuel consumption is not the largest amount as the acceleration is low at the 5th second.

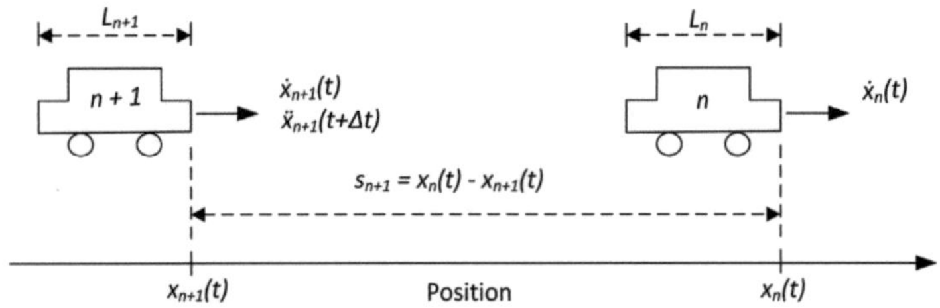

Fig. 4. Car-Following theory notations.

4. Traffic flow models

Traffic flow models are divided into macroscopic flow models and microscopic flow models. The macroscopic models measure a single value for the whole traffic flow (Chowdhury & Sadek, 2003). On the other hand, the microscopic models measure a single value for each vehicle (May, 1990).

Microscopic traffic flow models are well suited for ITS applications. These models are concerned with describing the flow by tracking individual vehicles instantaneously. The microscopic traffic flow models are either car-following or cellular automata.

4.1 Car-following models

Car-following models are time-continuous (May, 1990). All these models describe how one vehicle follows another vehicle. The car-following parameter is headway, which is applicable to individual pairs of vehicles within a traffic stream. Figure 4 shows a comprehensive set of car following theory notations. Definitions of these notations follow:

n: Leading vehicle;

$n+1$: Following vehicle;

L_n: Length of leading vehicle;

L_{n+1}: Length of following vehicle;

$x_n(t)$: Position of leading vehicle at time t;

$\dot{x}_n(t)$: Speed of leading vehicle at time t;

$\dot{x}_{n+1}(t)$: Speed of following vehicle at time t;

$\ddot{x}_{n+1}(t)$: Acceleration or deceleration rate of the following vehicle at time $t+\Delta t$;

Δt: Reaction time;

s_{n+1}: Space headway of following vehicle.;

The acceleration or deceleration rate occurs at time $t+\Delta t$. The reaction time is the time between t and the time the driver of the following vehicle decides to make an acceleration or deceleration. The time headway of the following vehicle can be determined as

$$h_{n+1} = s_{n+1}/\dot{x}_{n+1} \tag{2}$$

$(s_{i-1}\, s_i\, s_{i+1})_t$:	111	110	101	100	011	010	001	000
$(s_i)_{t+1}$:	1	0	1	1	1	0	0	0

Table 3. An example of CA rule table for updating the grid

$t = 0$:	1	0	1	0	1	0	1	0
$t = 1$:	0	1	0	1	0	1	0	1

Table 4. An example of grid configuration over one time step

$[\dot{x}_n(t) - \dot{x}_{n+1}(t)]$ is the relative speed of the leading vehicle and the following vehicle. The space headway will increase if the leading vehicle has a higher speed than the following vehicle. This implies that the relative speed is positive. On the other hand, if the relative speed is negative, the leading vehicle has lower speed than the following vehicle and the space headway is decreasing.

4.2 Cellular automata models

Cellular automata (CA) models are dynamic in which space and time are discrete. A cellular automaton consists of a grid of cells. Each cell can be in one of a finite number of states, which are updated synchronously in discrete time steps according to a rule. The rule is the same for each cell and does not change over time. Moreover, the rule is local which means the state of a cell is determined by the previous states of a surrounding neighborhood of cells. CA has been applied to study car traffic flow (Chopard et al., 2003; 1996). CA is simpler than car-following; however, it is less accurate and the locality of the rule makes drivers short-sighted, which means that they do not know if the leading vehicle will move or stop. Figure 5 shows the difference between space-continuous and space-discrete models.

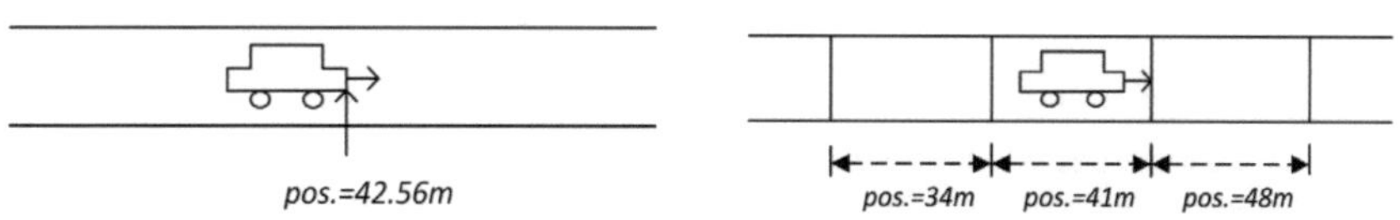

(a) Space-continuous (b) Space-discrete

Fig. 5. The difference between space-continuous and space-discrete models.

4.2.1 Example of a cellular automata model of car traffic

The model in this example is for a one-way street with one lane. The street is divided into cells. Each cell can be in one of two states (s). The first state represents an empty cell, denoted "0", while the second state represents a cell occupied by a vehicle, denoted "1". The movements of the vehicles are simulated as they jump from one cell to another ($i \rightarrow i+1$). The rule is that a vehicle jumps only if the next cell is empty. Consequently, the state of a cell is determined based on the states of its neighbors. In this model, each cell has two neighbors: one to its direct right, and one to its direct left. The car motion rule and the grid configuration over one time step can be explained as in Table 3 and Table 4, respectively.

The fraction of cars able to move is the number of motions divided by the total number of cars. For instance, in Table 4 at t=0, the number of motions is similar to the total number of

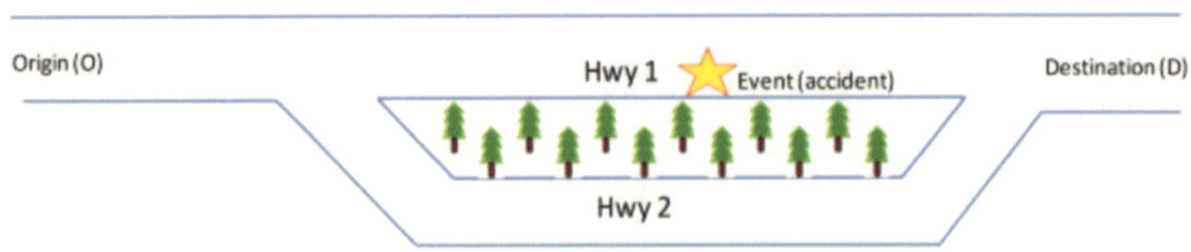

Fig. 6. Conceptional traffic model.

cars. They all equal to four. As a result, the fraction of cars that can move equals one. This indicates that the traffic is low in the system, and all the cars are able to move.

5. Geocast in vehicular networks for optimum reduction of vehicles' fuel consumption and emissions

By means of two examples, we show how vehicular networks can be used to reduce fuel consumption and carbon dioxide (CO_2) emission in a highway and a city environment. The first example is in a highway environment with the fuel consumption as the performance metric (Alsabaan et al., 2010a). The second example is in a city environment with considering the CO_2 emission as the performance metric (Alsabaan et al., 2010b).

5.1 Highway environment

Considering two highways (Hwys) and an accident occurred, this example illustrates the necessity of transmitting information to vehicles in order for drivers to choose the economical path. Simulation results demonstrate that significant amounts of fuel will be saved if such an economical geocast (EG) protocol is used.

5.1.1 System model

Since this work is quite interdisciplinary, models from different areas have to be considered. The system model includes (1) traffic model: represents the characteristics of the road network; (2) accident model: represents the characteristics of the accident; (3) fuel consumption and emission model: estimates the amount of fuel consumption and CO_2 emissions from vehicles; (4) communication model: represents the communication components and technologies that can be used for such an application.

Traffic Model: As shown in Figure 6, vehicles' trips initiate from the Original (O) to the Destination (D). Two Hwys with N-lanes have been considered. Hwy 1 with length $L\ 1$ is the main route for vehicles since it has the minimum travel time. Hwy 2 with length $L\ 2$, where $L\ 2 = 1.5 \times L\ 1$, is the alternative route. The free-flow speed of the highways is assumed to be 90 km/h.

Accident Model: An accident is modeled as temporal reductions in capacity, where such capacity reductions are specified as an effective number of lanes blocked by the accident for a given length and time. The model requires the following parameters:

- Start time of the accident;
- Time at which the traffic impact of the accident ends;
- Number of lanes blocked by the accident;
- Distance of the blocked lanes.

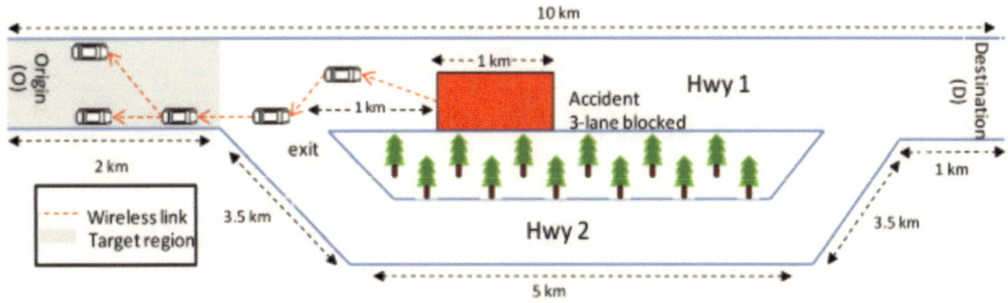

Fig. 7. System model.

Fuel Consumption and Emission Model: The VT-Micro model has been used in this example due to its simplicity and high accuracy. This model produces vehicle fuel consumption and emissions that are consistent with the ORNL data. The correlation coefficient between the ORNL data and the model predicted values ranges from 92% to 99% (Ahn et al., 2002). A more detailed description of the model is provided in Subsection 3.2.

Communication Model: Assume the existence of Inter-Vehicle Communication (IVC). Each vehicle is equipped with an Application Unit (AU) and On-Board Unit (OBU). It is assumed in this study that the AU can detect the crash occurrence of its vehicle. Moreover, it is assumed that the AU is equipped with a navigation system. It is also assumed in this example that the OBU is equipped with a (short range) wireless communication device. A multi-hop routing protocol is assumed in order to allow forwarding of data to the destination that has no direct connectivity with the source.

In this example, the use of geographical positions for addressing and routing of data packet (geocast) is assumed. The destination is addressed as all nodes in a geographical region. Designing or proposing the communication protocols that are suitable in applications such as reducing fuel consumption is out of the scope of this example. The main objective of this example is to encourage communications researchers to propose protocols with a goal of minimizing vehicle fuel consumption.

5.1.2 Simulation study

The lengths of the highways and accident are shown in Figure 7. Hwy 1 and Hwy 2 are 4-lane one direction. For both highways, the free-flow speed is 90 km/h. One of the important road segment characteristics is its basic saturation flow rate which is the maximum number of vehicles that would have passed the segment after one hour per lane. Another important characteristic is the speed at the basic saturation flow or speed-at-capacity. In this study, the basic saturation flow rate per lane is 2000 vehicles per hour with speed 70 km/h.

Vehicles enter the system uniformly in terms of the vehicle headway with a rate of 2500 vph/lane. For example, 2500 vehicles per hour uniformly depart from the origin between 9:00 and 9:10 am. In this case, a total of 864 vehicles will be generated with headway averaging 1.44 seconds.

The simulator used in this study is a trip-based microscopic traffic simulator, named INTEGRATION. The INTEGRATION model is designed to trace individual vehicle movements from a vehicle's origin to its destination at a deci-second level of resolution by

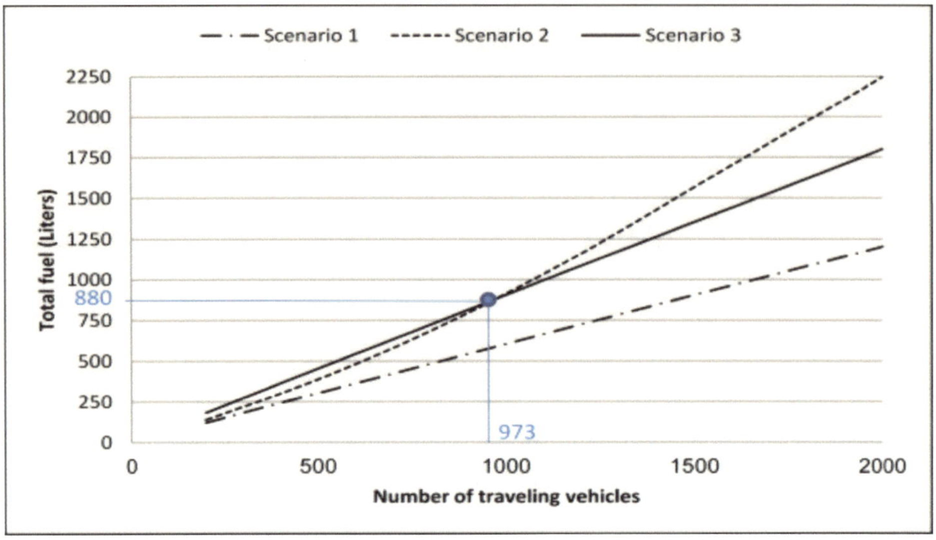

Fig. 8. Total fuel consumption versus number of traveling vehicles.

modeling car-following, lane changing, and gap acceptance behavior (Van, 2005a;b). In this paper, the total fuel consumption has been computed in four different scenarios:

Scenario 1: All vehicles traveled on Hwy 1 with no accident;

Scenario 2: All vehicles traveled on Hwy 1 where an accident is happened;

Scenario 3: All vehicles traveled on Hwy 2;

Scenario 4: Some vehicles changed their route from Hwy 1 to Hwy 2.

It is obvious that Hwy 1 is the best choice in terms of distance, travel time, fuel consumption, and emissions. However, if an accident happened on Hwy 1, it might not be the best choice for the drivers. Focusing on the fuel consumption, it is assumed that each vehicle has a navigation system that advises drivers on route selection based on minimizing trip fuel consumption. Figure 8 shows the impact of increasing the number of traveling vehicles on the total vehicles' fuel consumption. It is clear and expected that Scenario 1 is most economical. Consequently, the navigation system will advise the driver to travel on Hwy 1. However, if an accident happened on Hwy 1, a significant amount of fuel can be wasted due to stop-and-go conditions and congestion. It can be noticed from Figure 8 that Scenario 2 is more economical than Scenario 3 in light traffic density. Conversely, Scenario 3 becomes more economical than Scenario 2 with increasing traffic density.

Since navigation systems are not aware of the sudden events (e.g., accidents), vehicle-to-vehicle communications will be needed. With a focus on geocast, two main points have to be considered in order to design an economical protocol:

Geocast region The warning message has to be delivered to the region so that drivers can find a new path to avoid congestion.

Delivered Message A warning message will be issued once an accident occurs in order to alert nearby vehicles. Based on the results shown in Figure 8, not all alert routes (i.e. routes with accident "Scenario 2") consume more fuel than no alert routes "Scenario 3". Therefore, we need to define when the status of the most economical route will change from Hwy 1 to Hwy 2 (this depends on the traffic density). Then, find a way to inform the drivers.

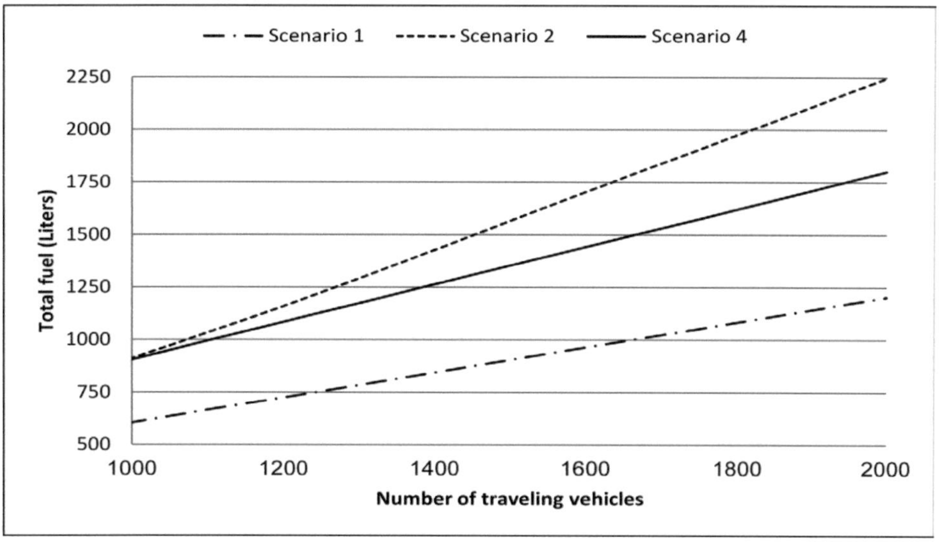

Fig. 9. The impact of the EG protocol on the amount of fuel consumption.

The presented system model requires a geocast protocol that can inform all vehicles, which have traveled beyond the nearest exit to the accident site. Moreover, it needs a geocast protocol that is able to advise the first 973 traveling vehicles to continue on Hwy 1, while others change their route to Hwy 2. Figure 9 shows the amount of fuel consumption if the above requirements can be met.

5.1.3 Discussions

The important issues that have to be taken into account in designing an economical geocast protocol in this model are as follows:

Calculating the Desirable Number of Traveling Vehicles on Hwy 1: The desirable number of traveling vehicles is 973 in this study. This number was obtained from the simulator. However, research is needed to be done to estimate this number. This information can be used in designing communication applications. Consequently, the geocast packet will contain this information when a geocast is performed. In conclusion, ITS applications and tools should be able to calculate this kind of information and inject it to the geocast packet.

Traffic Density versus Fuel Consumption: In many cases the shortest path in terms of time or distance will also be the minimum fuel consumption. However, this is not true in several cases which increase traffic. For instance, congestion will start if an accident happened. In this case, stop-and-go conditions will occur; thus, more fuel will be consumed. Therefore, changing to another path even if it is longer is preferred. In addition, it is important to point that in some cases, an accident might happen on a highway, but the vehicles do not need to change the path since it is still the best in terms of fuel consumption. This issue depends on the traffic density.

Defining the region of interest: In this work, the target region is 2 km beyond the nearest exit. However, the idea of region of interest needs to be investigated. In references (Rezaei et al., 2009a;b; Rezaei, 2009c), the region of interest has been determined base on the type of warning messages and traffic density. Moreover, two metrics have been defined to

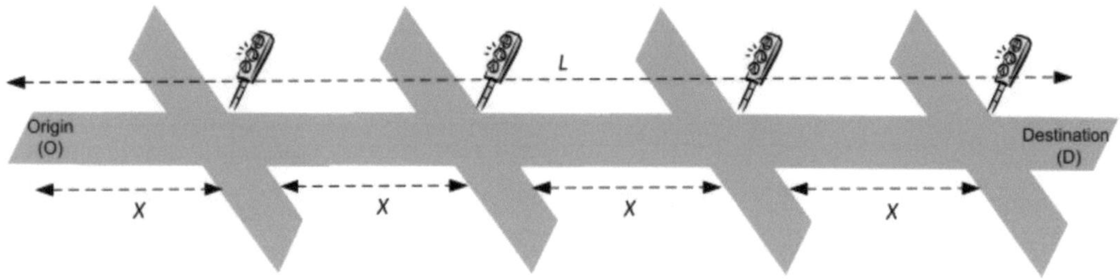

Fig. 10. Conceptual traffic model.

study the effect of data dissemination: communication cost and additional travel cost. Communication cost is the least number of vehicles involved in retransmitting, while the additional travel cost is the cost differences associated with the paths calculated before and after propagating information.

5.2 City environment

This example illustrates the benefit of transmitting the traffic light signal information to vehicles for CO_2 emission reduction. Simulation results demonstrate that vehicle CO_2 emission will be reduced if such an environmentally friendly geocast (EFG)protocol is used.

5.2.1 System model

Traffic Model: As shown in Fig. 10, a vehicle's trip initiates from the Origin (O) to the Destination (D). A street segment with length L and N-lanes has been considered. This segment has four static Traffic Light Signals (TLSs). The distance between each TLS and the following one is X. Each TLS has three phases: green, yellow, and red. The phase duration is T_g, T_y, and T_r for green, yellow, and red, respectively. The free-flow speed of the street is S_F m/s.

Fuel Consumption and Emission Model: Similar to the example in Section 5.1, the VT-Micro model has been used in this example due to its simplicity and high accuracy.

Communication Model: Assume that TLSs and the traveling vehicle are equipped with an On-Board Unit (OBU). The assumption that the OBU is equipped with a (short range) wireless communication device is considered. In addition to OBU, the traveling vehicle is equipped with an Application Unit (AU). It is assumed that the AU is equipped with position data and map (e.g., GPS). Therefore, the vehicle knows its location and the location of TLSs. The TLSs are the transmitters, while the destination is addressed as all vehicles in a Region of Interest (ROI).

Each TLS sends a geocast packet within a transmission range which is equal to the ROI. This packet is directed to vehicles approaching the signal. The geocast packet is considered to contain three types of information:

1. Type of the current phase (either green "g", yellow "y", or red "r");
2. Number of seconds to switch from the current phase (L_g, L_y, or L_r);
3. Traffic light schedule, which includes the full green, yellow, and red phase time (T_g, T_y, and T_r).

With these information, the vehicle calculates a recommended speed (S_R) for the driver to avoid stopping at the TLS. S_R can be calculated as the distance between the vehicle and the

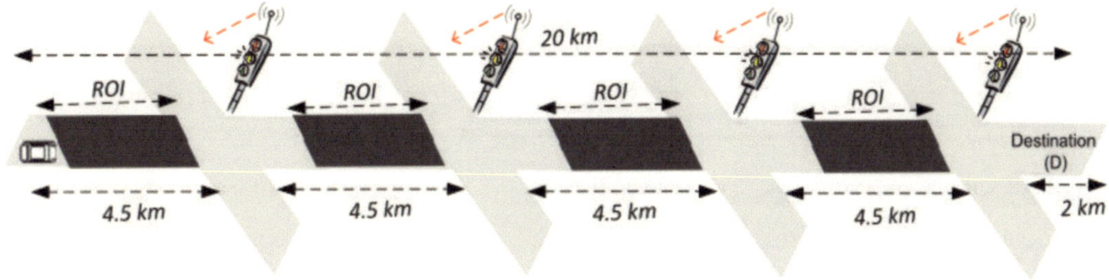

Fig. 11. System model.

TLS after receiving the packet (d) over the required delay of the vehicle to be able to pass the TLS. Calculating this delay depends on d and the information in the geocast packet. The maximum allowed speed for vehicles equals S_F. The following equations show the calculation of S_R:

- If the current phase is green

$$S_R = \begin{cases} S_F, \text{ if } d/S_F \leq L_g \\ \min(\max(\frac{d}{(N_g-1)\cdot C_L+L_g+T_y+T_r+M-D}, S_{R_{min}}), S_F), \\ \text{otherwise} \end{cases} \quad (3)$$

where $N_g = \lceil \frac{d/S_F - L_g}{C_L} \rceil$

- If the current phase is red

$$S_R = \begin{cases} S_F, \text{ if } L_r < d/S_F \leq L_r + T_g \\ \min(\max(\frac{d}{N_r\cdot C_L+L_r+M-D}, S_{R_{min}}), S_F), \quad \text{otherwise} \end{cases} \quad (4)$$

where $N_r = \lceil \frac{d/S_F - L_r - T_g}{C_L} \rceil$

- If the current phase is yellow

$$S_R = \begin{cases} S_F, \text{ if } T_r + L_y < d/S_F \leq T_r + L_y + T_g \\ \min(\max(\frac{d}{N_y\cdot C_L+T_r+L_y+M-D}, S_{R_{min}}), S_F), \quad \text{otherwise} \end{cases} \quad (5)$$

where $N_y = \lceil \frac{d/S_F - T_r - L_y - T_g}{C_L} \rceil$, $C_L = T_g + T_y + T_r$, C_L is the TLS cycle length, $S_{R_{min}}$ is the minimum recommended speed (m/s). N_g, N_r, and N_y represent the number of light cycles completed before the vehicle can pass the TLS when the current phase is green, red, and yellow, respectively. D is the packet delay (s) and M is a margin value (s).

Margin value is the number of seconds that represent the sum of the time the vehicle has to comfortably decelerate from its current speed to the recommended speed and the time the vehicle has to decelerate when it approaches a red TLS.

5.2.2 Simulation study and discussions

The length of the street and the distances between the TLSs are shown in Figure 11. The street has one lane and is in one direction. The ROI is changeable from 0.5 to 4.5 km in increments of 0.5. The rest of the simulation parameters are specified in Table 5.

S_F	60 km/h	T_y	5 s
$S_{R_{min}}$	40 km/h	T_r	50 s
C_L	100 s	D	0 s
T_g	45 s	M	10 s

Table 5. Simulation Parameters

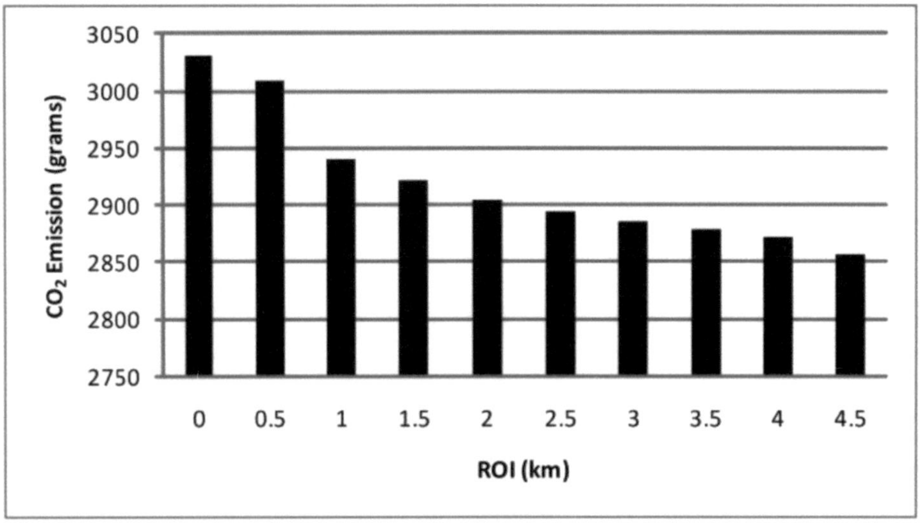

Fig. 12. Vehicle CO_2 emission versus region of Interest.

The simulator used is INTEGRATION. In this example, the total vehicle's CO_2 emission have been computed at different ROIs. We assume that the TLSs send a packet at the moment when the vehicle entered the ROI. In this case, the distance between the vehicle and the TLS is almost equal to the ROI.

Figure 12 shows the impact of the length of the ROI on the amount of the vehicle's CO_2 emission. With a large ROI, the vehicle will have more time to avoid stops and accelerations. Therefore, the amount of the vehicle's CO_2 emission decreases with increasing ROI as shown in the figure.

Figure 13 shows how stopping time decreases as ROI increases for a vehicle that travels from O to D. It is clear that in the absence of communication between the TLSs and the vehicle, the vehicle will stop for around 75 seconds. This time would be shortened if the idea of vehicular networks is applied. It can be seen that the vehicle will keep passing all TLSs without stopping when the geocast packet can cover at least 1 km ahead of each TLS.

Consider that the vehicle travels at the free-flow speed if it is out of the ROI. After receiving the geocast packet from the TLS, the vehicle will recommend to the driver the environmentally friendly speed as in Eqs. 3, 4, and 5. The goal of calculating this speed is to have the vehicle avoid unnecessary stops, useless acceleration and high speed.

The vehicle may avoid a stop by adapting its speed to (S_R), such that $S_{R_{min}} \leq S_R \leq S_F$. The vehicle will adjust its speed to $S_{R_{min}}$ in order to avoid useless high speed if it is impossible for the vehicle to avoid stopping. The last goal is to alleviate vehicle accelerations. This can be

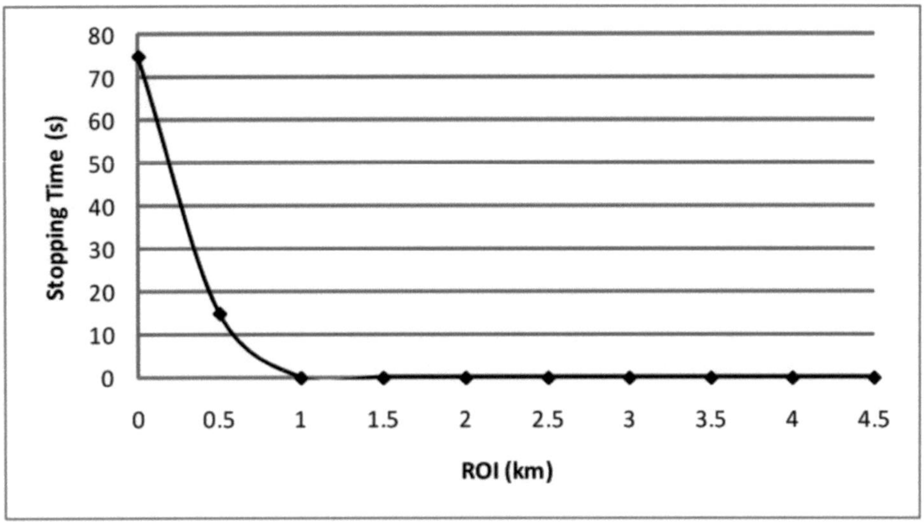

Fig. 13. Vehicle stops delay versus region of Interest.

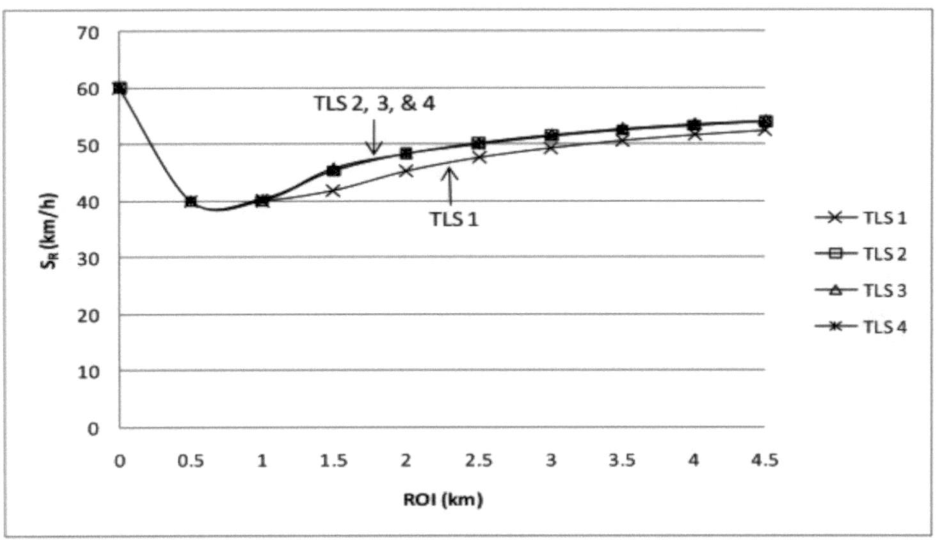

Fig. 14. Recommended speed versus region of Interest.

achieved by calculating the S_R as the maximum possible speed for the vehicle to pass the TLS with no stops. As a result, after passing the TLS, the vehicle will return to the free-flow speed with low acceleration.

Figure 14 shows the impact of the increase of the ROI on the S_R. With no vehicular network, the vehicle is not aware of the TLS information; therefore, it travels at the maximum allowed speed. At ROI = 0.5 km, the vehicle will realize that stopping will happen. Consequently, the recommended speed is reduced to SR_{min}. After that, the S_R will increase with increasing ROI.

Figure 15 shows the benefit of increasing the ROI to alleviate average vehicle acceleration. At ROI = 0 and 0.5 km, the vehicle stops at each TLS. Therefore, the average vehicle acceleration

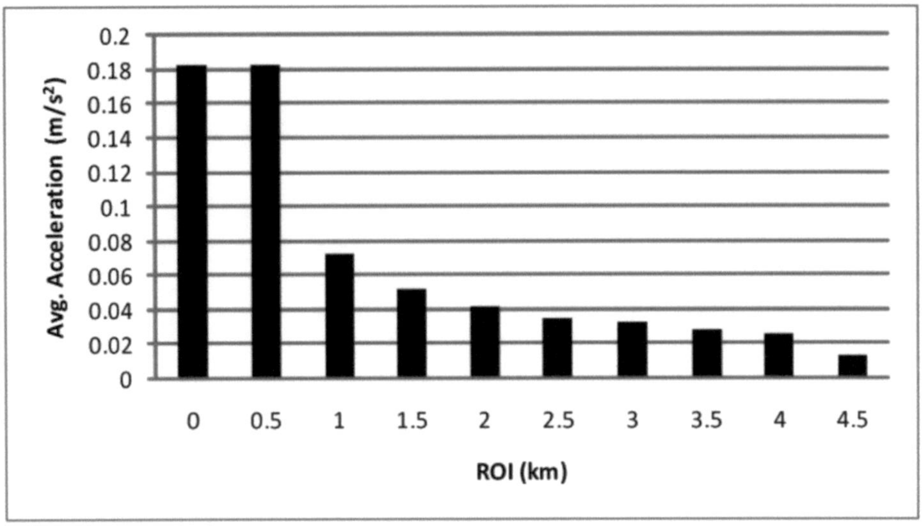

Fig. 15. Average acceleration versus region of Interest.

at ROI = 0 and 0.5 km are the same. However, the CO_2 emission are less at ROI = 0.5 km as shown in Figure 12. This is because the recommended speed at ROI = 0.5 km is reduced to SR_{min} in order to avoid useless high speed.

6. Conclusions

This chapter is to show the impact of vehicular networks on vehicle fuel consumption and CO_2 emissions. This chapter also aims to motivate researchers working in the field of communication to design EEFG protocols, and demonstrate the ability to integrate fuel consumption and emission models with vehicular networks. The first example was in a highway environment with the fuel consumption as the performance metric. This example illustrates the necessity of sending information to vehicles in order for drivers to choose an appropriate path to a target to minimize fuel consumption. Simulation results demonstrate that significant amounts of fuel will be saved if such an EG protocol is used. The second example was in a city environment with considering the CO_2 emission as the performance metric. This example illustrates the benefit of transmitting the traffic light signal information to vehicles for fuel consumption and emission reduction. Simulation results demonstrate that vehicle fuel consumption and CO_2 emissions will be reduced if such an environmentally friendly geocast protocol is used.

7. Recommendations for future work

A suggested future research is to develop a communication protocol that considers the multidisciplinary research area in order to reduce vehicle fuel consumption and CO_2 emissions. This protocol should be able to deal with different traffic scenarios and events such as accidents and congestion. Another future work is to consider the electric vehicles. In this case, the goal will be how to apply vehicular networks in order to reduce the battery energy consumption.

8. References

Alsabaan, M.; Naik, K.; Nayak, A. (2010a). Applying Vehicular Ad Hoc Networks for Reduced Vehicle Fuel Consumption. *The* 2nd *International Conference on Wireless & Mobile Networks (WiMo)*, pp. 217–228.

Alsabaan, M.; Naik, K.; Khalifa, T.; Nayak, A. (2010b). Vehicular Networks for Reduction of Fuel Consumption and CO_2 Emission. *The* 8th *IEEE International Conference on Industrial Informatics (INDIN)*, pp. 671–676.

Ahn, K. & Rakha, H. (2007). Field Evaluation of Energy and Environmental Impacts of Driver Route Choice Decisions, *Proceedings of the IEEE Intelligent Transportation Systems Conference*, pp.730–735.

Ahn, K.; Rakha, H.; Trani, A.; Van Aerde, M. (2002). Estimating vehicle fuel consumption and emissions based on instantaneous speed and acceleration levels, *Journal of Transportation Engineering*, Vol. 128, No.2, pp. 182–190.

Barth, M.; Boriboonsomsin, K.; Vu, A. (2007). Environmentally-friendly navigation, *Proceedings of the IEEE Intelligent Transportation Systems Conference*, pp. 684–689.

Barth, M.; An, F.; Younglove, T.; Scora, G.; Levine, C.; Ross, M.; Wenzel, T. (2000). Comprehensive modal emission model (CMEM), version 2.0 user's guide. Riverside, California.

Broustis, I. & Faloutsos, M. (2008). Routing in Vehicular Networks: Feasibility, Modeling, and Security, *International Journal of Vehiculer Technology*, Vol. 2008, Article ID 267513.

Briesemeister,L.; Schafers, L.; Hommel, G. (2000). Disseminating messages among highly mobile hosts based on inter-vehicle communication, *IEEE IV*, pp. 522–527.

Chowdhury, M.A. & Sadek, A.W. (2003). Fundamentals of intelligent transportation systems planning, *Artech House Publishers*.

California Air Resources Board (2007). User's Guide to EMFAC, Calculating emission inventories for vehicles in California.

Chopard, B.; Dupuis, A.; Luthi, P.O. (2003). A cellular automata model for urban traffic and its application to the city of Geneva, *Network-Spatial-Theory*, Vol. 3, pp. 9–21.

Chopard, B.; Luthi, P.O.; Queloz,P-A. (1996). Cellular automata model of car traffic in a two-dimensional street network, *Journal of Physics A: Mathematical and General*, Vol. 29, No. 10, pp. 2325–2336.

Environmental Protection Agency (2002). User's Guide to Mobile 6, Mobile Source Emission Factor Model, Ann Arbor, Michigan.

Kuriyama, H.; Murata, Y.; Shibata, N.; Yasumoto, K.; Ito, M. (2007). Congestion Alleviation Scheduling Technique for Car Drivers Based on Prediction of Future Congestion on Roads and Spots, *Proceedings of the IEEE Intelligent Transportation Systems Conference*, pp. 910–915.

Karp, B. & Kung, H.T. (2000). GPSR: greedy perimeter stateless routing for wireless networks, *Proc. MobiCom*, pp. 243–254.

Ko, Y.B. & Vaidya, N.H. (2000). Location-Aided Routing (LAR) in mobile ad hoc networks, *Wireless Networks*, Vol. 6, No. 4, pp. 307–321.

Krajzewicz, D.; Hertkorn, G.; Wagner, P.; Rössel, C. (2002). SUMO (Simulation of Urban MObility): An open-source traffic simulation, *Proceedings of the* 4th *Middle East Symposium on Simulation and Modelling*, pp. 183–87.

Moustafa, H. & Zhang, Y. (2009). Vehicular Networks: Techniques, Standards, and Applications, *Auerbach Publications*, Taylor & Francis Group, USA.

Maihöfer, C. (2004). A Survey of Geocast Routing Protocols, *IEEE Communications Surveys and Tutorials*, Vol. 6, No. 2, pp. 32–42.

Moreno, M.T. (2007). Inter-Vehicle Communications: Achieving Safety in a Distributed Wireless Environment: Challenges, Systems and Protocols, *Dissertation*, ISBN: 978-3-86644-175-0, Universitätsverlag Karlsruhe.

Maihofer, C. & Eberhardt, R. (2004). Geocast in vehicular environments: caching and transmission range control for improved efficiency, *IEEE IV*, pp. 951–956.

May, A.D. (1990). *Traffic flow fundamentals*, Prentice Hall.

Organization for Economic Co-operation and Development (OECD)/ International Energy Agency (IEA) (2009). CO_2 Emissions from Fuel Combustion Highlights

Rezaei, F.; Naik, K.; Nayak, A.(2009a). Investigation of Effective Region for Data Dissemination in Road Networks Using Vehicular Ad hoc Network, *IEEE International Conference on Fuzzy Systems*, Korea.

Rezaei, F.; Naik, K.; Nayak, A. (2009b). Propagation of Traffic Related Information in Road Networks, *Canadian Society for Civil Engineering*.

Rezaei, F. (2009c). Investigation of Effective Region for Warning Data Dissemination in Vehicular Networks, *A thesis Presented to the University of Waterloo*, Department of Electrical and Computer Engineering, Waterloo, Ontario, Canada.

Sichitiu, M.L. & Kihl, M. (2008). Inter-vehicle communication systems: a survey, *IEEE Communications Surveys & Tutorials*, Vol. 10, Iss. 2, pp. 88–105.

Tsugawa,S. & Kato, S.(2010). Energy ITS: another application of vehicular communications, *IEEE Communications Magazine*, Vol. 6, No. 11, pp. 120–126.

UK Department for Environment (2008). Food & Rural Affairs: Cold Start Advanced-user guide. iss. 1

Van Aerde, M. & Associates, Ltd. (2005a). INTEGRATION release 2.30 for Windows: User's guide – Volume I: Fundamental features.

Van Aerde, M. & Associates, Ltd. (2005b). INTEGRATION release 2.30 for Windows: User's guide – Volume II: Fundamental features.

Wiebe, E.C.(2011). Gasoline Prices in Parts of Canada between 1998 and 2011, `http://climate.uvic.ca/people/ewiebe/car/fuel_price.html`

2

An Investigation of Measurement for Travel Time Variability

Steven Chien[1] and Xiaobo Liu[2,*]
[1]*New Jersey Institute of Technology, Newark, NJ*
[2]*Jacobs Engineering Group, Morristown, NJ*
USA

1. Introduction

Congestion has grown over the past two decades making the travel time highly unreliable. The Federal Highway Administration (FHWA), US Department of Transportation (USDOT) indicated travel time as an important index to measure congestion. Frequent but stochastic, irregular delays increase the challenge for people to plan their journey - e.g. when to depart from the origin, which mode(s) and route(s) to use so the on-time arrival at the destination can be ensured. In addition to average travel time, the reliability of travel time has been deemed as an index for quantifying the effects of congestion, which can be applied to the areas of transportation system planning, management and operations as well as network modeling. Travel time reliability has been classified into two categories: probability of a non-failure over time and variability of travel time (Elefteriadou and Cui, 2005).

The variability of travel time plays an important role in the Intelligent Transportation Systems (ITS) applications. According to the Safe, Accountable Flexible, Efficient Transportation Equity Act: A Legacy for Users (SAFETEA-LU) Reporting and Evaluation Guidelines, travel time variability indicates the variability of travel time from an origin to a destination in the transportation network, including any modal transfers or en-route stops. This measure can readily be applied to intermodal freight (goods) movement as well as personal travel. Reducing the variability of travel time increases predictability for which important planning and scheduling decisions can be made by travelers or transportation service suppliers. In the advent of ITS, the incident impact, such as delay, may be reduced via disseminating real-time traffic information, such as the travel time and its variability. Travel time variability has been extensively applied in transportation network models and algorithms for finding optimal paths (Zhou 2008). The recent ITS applications highlighted the needs for better models in handling behavioral processes involved in travel decisions. It was indicated that travel time variability has affected the travelers' route choice (Avineri and Prashker, 2002). The relationship between the estimated travel time reliability and the frequency of probe vehicles was investigated by Yamamoto et al (2006). It was found that the accuracy of travel time estimates using low-frequency floating car data (FCD) appears little different from high-frequency data.

---

*Corresponding Author

Recker et al. (2005) indicated that travel time variability is increasingly being recognized as a major factor influencing travel decisions and, consequently, is an important performance measure in transportation management. An analysis of segment travel time variability was conducted, in which a GIS traffic database was applied. Standard deviation and normalized standard deviation were used as measures of variability. Brownstone et al. (2005) indicated that the most important facor is the "value of time" (VOT), i.e. the marginal rate of travel time substitution for money in a travelers' indirect utility function. Another factor is the value of reliability (VOR), which measures travelers' willingness to pay for reductions in the day-to-day variability of travel times facing a particular type of trip. Bartin and Ozbay (2006) identified the optimal routes for real-time traveler information on New Jersey Turnpike, which maximizes the benefit of motorists. The variance of travel times within a time period over consecutive days was employed as an indicator of uncertainty. With the concept of multi-objective approach, Sen and Pillai (2001) developed a mean-variance model for optimizing route guidance problems. The tradeoff between the mean and variability of travel time was discussed. For improving decision reliability, Lu et al. (2005) developed a statistic method to analyze the moments and central moments of historic travel time data, which provided quantitative information on the variability and asymmetry of travel time. Palma et al. (2005) conducted a study in Paris to determine the route choice behavior when travel time is uncertain. Both the mean and variability of travel time were considered.

2. Objective

The objective of this chapter is to investigate the measurement of travel time variability and reliability with FCD. Considering the Variability of Travel Time (VTT) as a component of mobility performance metrics, this chapter discusses technologies and methodology applied to collect, process and analyze the travel time data. To analyze the impact of travel time due to non-recurring congestion, three case studies on selected highways were conducted. As defined in a report titled "Traffic Congestion and Reliability: Trends and Advanced Strategies for Congestion Mitigation" (FHWA, 2005), the travel time reliability was deemed as how much travel time varies over the course of time. The variation in travel times from one day to the next is due to the fact that underlying conditions (such as vehicle composition, weather conditions) vary widely. Seven sources of congestion are identified, including physical bottlenecks ("capacity"), traffic incidents, work zones, weather, traffic control devices, special events, and fluctuations in normal traffic condition, which contribute to total congestion and conspire to produce biased travel time estimates.

3. Statistical indices

To estimate travel time variability with FCD, statistical formulas for generating suitable reliability indicators, such as mean, standard deviation, the 95th percentile travel time, and buffer index, etc are utilized.

Mean Travel Time (Tl)

The mean travel time, denoted as T_l, is equal to the sum of the travel time collected by a number of floating cars, denoted as n, traveling on Link l. Thus,

$$T_l = \frac{1}{\mathrm{n}}\sum_{i=1}^{n} \mathrm{t}_{\mathrm{li}} \qquad \forall\, \mathrm{l} \tag{1}$$

where t_{li} represents the travel time of the ith vehicle spent on Link l.

Standard Deviation of Travel Time (σ_l)

The standard deviation of travel time on Link l, denoted as σ_l, is the measure of the dispersion of travel times, which can be formulated as

$$\sigma_l = \sqrt{\frac{\sum_{i=1}^{n}(t_{li} - T_l)^2}{n-1}} \qquad \forall l \qquad (2)$$

The 95th Percentile Travel Time ($T_{95\% l}$)

Travel time reliability may be measured by using the percentile travel time, which indicates the delay on a particular link. The 95th percentile travel time of Link l, denoted as $T_{95\% l}$, is the travel time of which 95% of sample travel times are at or below this amount. The difference between the mean (T_l) and the 95th percentile travel time ($T_{95\% l}$) is called buffer time, denoted as T_{Bl}, which also represents the extra time needed to compensate for unexpected delays.

Buffer Index (B_l)

The buffer index of Link l, denoted as B_l, may be applied to estimate the extra time that travelers should add onto the mean travel time to ensure on-time or earlier arrivals. The buffer index, denoted as p, is in percent, which increases as the reliability of travel time gets worse. For example, for the mean travel time T_l of Link l with a buffer index p means a traveler should reserve additional travel time pT_l, called buffer time, to ensure on-time arrival at the destination. The buffer index B_l is defined as the 95th percentile travel time ($T_{95\% l}$) minus the mean travel time (T_l) divided by the mean travel time. Thus,

$$B_l = \left(\frac{T_{95\% l} - T_l}{T_l}\right)(100\%) \qquad \forall l \qquad (3)$$

Other percentile travel times (e.g. the 85th, 90th, or 99th percentiles) could be used depending upon the desired level of reliability. A lower percentile travel time was suggested to calculate reliability measures for less critical routes.

Planning Time Index (PI)

The planning time index is introduced to estimate total travel time when an adequate buffer time is included, which differs from the buffer index which considers recurring and non-recurring delays. Thus, the planning time index compares the longest travel time against a travel time incurred by free-flow traffic. For the travel time consumed under free flow condition on Link l, denoted as TFl, a planning time index P_l indicates that the extra time should be planned is the product of T_{Fl} and P_l. The planning time index in percent may be obtained from the 95th percentile travel time divided by the free-flow travel time and multiplied 100%. Thus,

$$P_l = \left(\frac{T_{95\% l}}{T_{Fl}}\right)(100\%) \qquad \forall l \qquad (4)$$

The planning time index is useful since it can be directly compared to the travel time index (a measure of average congestion) on similar numeric scales. Note that the buffer index represents the additional percent of travel time that is necessary above the mean travel time, whereas the planning time index represents the total travel time that is necessary. Thus, the buffer index was used to be applied to estimate non-recurring delays.

Various travel time reliability measures have been discussed in previous studies, but only few of them are effective in terms of communication with road users and general public. Buffer index (BI) and planning time index (PI) are ones of the most effective methods in measuring travel time reliability (FHWA). Other statistical measures, such as standard deviation and coefficient of variation, have been used to quantify travel time reliability, but they are not easy for a non-technical audience to understand and would be less-effective communication tools.

4. Case studies

With FCD, three case studies are conducted for investigating the variability of travel time, including (1) analysis of temporal and spatial travel times, (2) evaluation of adverse weather impact to travel time variability and reliability, and (3) investigating travel time variability in a freeway network.

4.1 Case study I – Analysis of temporal and spatial travel times

The first set of the travel time FCD were collected to assess the impacts of special events, traffic control devices and bottleneck on travel time variability. The travel time data were collected by floating cars with GPS-based sensors traveling on the segments of I-295 and the New Jersey Turnpike (NJTP) from 10:00 to 19:00 on three different Sundays: May 24, June 7, and July 19 in 2009. On an hourly basis starting at 10:00, two floating cars were dispatched simultaneously, one heading for the I-295 segment and the other heading for the NJTP segment. To investigate the travel time variability and congestion of the studied segments, the collected data were analyzed on a link (e.g., links 1, 2, and 3) and a path (e.g., from Node 1 to Node 4) basis, as shown in Figure 1. The geographic information of each route and all their links is summarized in Table 1.

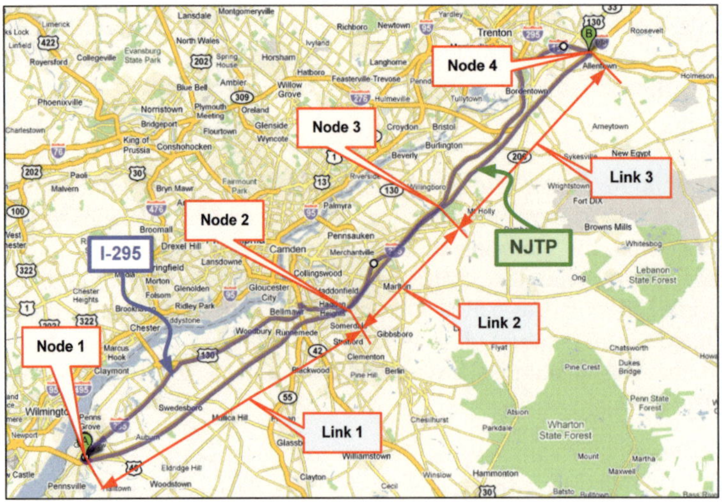

Fig. 1. Study Segments on I-295 and NJTP.

Route Name	Node	Latitude	Longitude	Description	Link Distance (mi) (mile)	
NJTP (A)[1]	1	39.681461	-75.478286	Pilot Truck Stop	Link 1	28.6
	2	39.873189	-75.017678	near Excess Rd	Link 2	11.6
	3	39.994728	-74.867119	near Creek Rd	Link 3	19.9
	4	40.194372	-74.608439	NJTP Exit 7-A Toll	Total	60.1
NJTP (B)[2]	1	39.681411	-75.468467	NJTP Entry 1 Toll	Link 1	27.9
	2	39.873189	-75.017678	near Excess Rd	Link 2	11.6
	3	39.994728	-74.867119	near Creek Rd	Link 3	19.4
	4	40.190761	-74.605322	NJTP Off-Ramp to Exit 7-A	Total	58.9
I-295	1	39.689767	-75.473067	Pilot Truck Stop	Link 1	30.5
	2	39.873944	-75.018075	near Excess Rd	Link 2	11.4
	3	39.995669	-74.868278	near Creek Rd	Link 3	22.5
	4	40.197547	-74.613300	NJTP Exit 7-A	Total	64.4

1: Segment between interchanges of Rt. 140/NJTP (near Delaware Memorial Bridge) and NJTP Exit 7-A
2: Segment between the toll facility at the beginning entry point of NJTP and the beginning of off-ramp to Exit 7-A

Table 1. Node Locations of I-295 and NJTP (A) and (B).

Since tolls are collected at the entrances and exits of NJTP, the travel times of two segments [e.g., NJTP (A) and NJTP (B)] were prepared and estimated. NJTP (A) includes the toll payment process and starts from interchange Rt. 140/NJTP and ends at NJTP Exit 7-A. To estimate the travel time excluding delays at toll plazas, NJTP (B) starts right after the first toll facility on the NJTP and ends at the beginning of the off ramp to Exit 7-A. The impact of tolls on the travel time was estimated by comparing the travel time difference between these segments as shown in Table 2. It can be easily observed that the first toll facility contributes 1.6 minutes average delay and the second toll facility provides a slightly longer average delay as 1.8 minutes. The standard deviation to the travel time is 0.3 minute for the first toll and 0.5 minute for the second toll based on a 9-hour observation period from 10:00 to 19:00.

The link- based travel times are collected by probe vehicles dispatched at each hour on survey days. Table 3 summarizes the mean travel times and the associated standard deviations. The speeds are also estimated based on the travel time and the corresponding link distance, and the statistical results are summarized in Table 4. It can be observed in Table 3 that less travel time was consumed on Link 1 of I-295 than that of NJTP with standard deviation to the mean. For Link 2, the mean travel times of I-295 and NJTP were the same, but the standard deviations to the means of NJTP for all travel time data on the three survey days were consistently less than that of I-295, which indicates that Link 2 of NJTP is more reliable than that of I-295. For Link 3, the mean travel time of NJTP was 20.8 minutes, slightly shorter than that of I-295 (e.g. 21.6 minutes). The standard deviation of travel time of NJTP was significantly higher than that of I-295, was incurred by congestion on NJTP (See Figures 2 and 3).

Departure Time	24-May-09 Time Difference (minutes)			7-Jun-09 Time Difference (minutes)			19-Jul-09 Time Difference (minutes)		
	Link1	Link3	Path	Link1	Link3	Path	Link1	Link3	Path
10:00	1.6	2	3.6	2	2	4	2	2	4
11:00	1.4	1.5	2.9	1.7	1.5	3.1	1.1	1.1	2.2
12:00	1.3	2.9	4.2	2	1.9	3.8	1.1	2	3.1
13:00	1.1	1.4	2.5	1.9	1.4	3.2	1.1	1.4	2.5
14:00	1.8	1.6	3.4	2.3	1.6	3.8	1.6	2.3	3.9
15:00	1.8	1.5	3.3	1.3	2.1	3.4	1.4	1.4	2.8
16:00	1.2	1.8	3	1.8	1.8	3.7	1.8	2.5	4.3
17:00	2	1.7	3.7	1.4	1.7	3.1	1.4	2.5	3.9
18:00	2	1.3	3.3	1.8	1.3	3.2	1.8	3.3	5.1
19:00	1.7	1.4	3.1	2	1.9	3.9	1.1	1.4	2.5
Mean TT	1.6	1.7	3.3	1.8	1.7	3.5	1.4	2.0	3.4
STD TT	0.3	0.5	0.5	0.3	0.3	0.4	0.3	0.7	0.9

Table 2. Travel Time Difference between NJTP (A) and NJTP (B).

Mean (Minutes)	NJTP				I-295			
Dates	Day 1	Day 2	Day 3	3-days	Day 1	Day 2	Day 3	3-days
Link 1	28.1	28.1	28.1	28.1	27	26.4	25.9	26.5
Link 2	10.9	10.4	10.6	10.6	10.8	11.0	10.0	10.6
Link 3	17.4	19.5	25.5	20.8	21.2	21.3	22.2	21.6
STD (Minutes)	NJTP				I-295			
Dates	Day 1	Day 2	Day 3	3-days	Day 1	Day 2	Day 3	3-days
Link 1	0.9	1.3	2.2	1.50	1.1	1.2	2.6	1.79
Link 2	0.8	0.3	1.0	0.77	0.4	1.1	1.2	1.05
Link 3	0.8	4.6	6.3	5.60	0.8	1.2	3.1	1.97

1: Day 1, 2 and 3 represents 5/24/09, 6/7/09, and 7/19/09, respectively.

Table 3. Mean and Standard Deviation of Travel Times along NJTP (A) and I-295.

Mean (mph)	NJTP				I-295			
Dates	Day 1	Day 2	Day 3	3-days	Day 1	Day 2	Day 3	3-days
Link 1	61.1	61.2	61.3	61.2	67.9	69.2	71.4	69.5
Link 2	64.0	67.3	66.4	65.9	63.5	62.8	69.3	65.3
Link 3	68.7	63.7	49.3	60.6	63.7	63.4	61.9	63.0
STD (mph)	NJTP				I-295			
Dates	Day 1	Day 2	Day 3	3-days	Day 1	Day 2	Day 3	3-days
Link 1	1.9	2.7	4.7	3.20	2.7	3.2	7.6	5.08
Link 2	4.5	1.9	5.7	4.40	2.6	5.7	8.9	6.77
Link 3	3.0	11.6	11.4	12.45	2.3	3.5	8.0	5.12

Table 4. The Mean and Standard Deviation of Travel Speed along NJTP (A) and I-295.

Figure 2 illustrates the path travel time variation at different departure times on July 19, 2009. In Figure 2, the travel times on NJTP (A) and NJTP (B) considerably varied over time, which were longer than that of I-295 between 13:00 and 19:00.

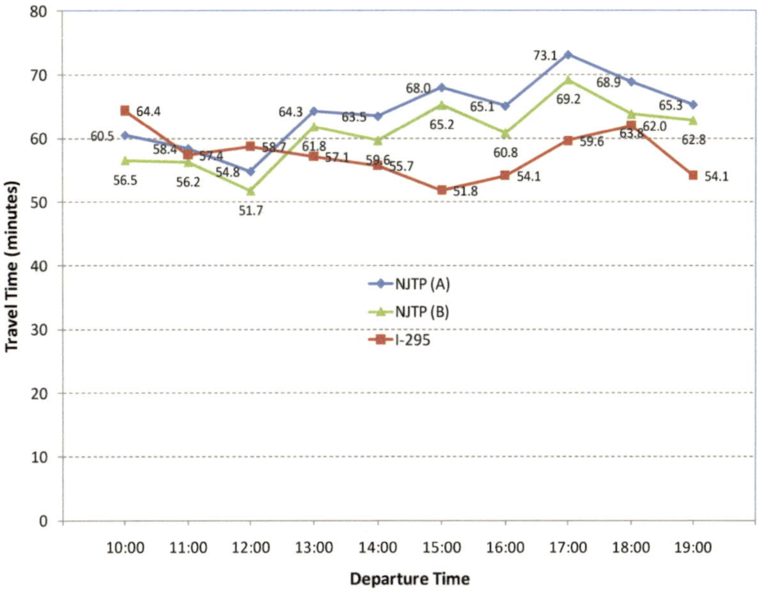

Fig. 2. Travel Time vs. Departure Time (July 19, 2009).

All vehicle travel profiles on NJTP (A) are illustrated in Figure 3 where the slope of each vehicle profile represents speed. The speed before entering the toll facility at the entrance of the NJTP was found generally low. As highlighted in purple, the probe vehicle which departed at 13:00 experienced congestion over the entire segment compared with vehicles

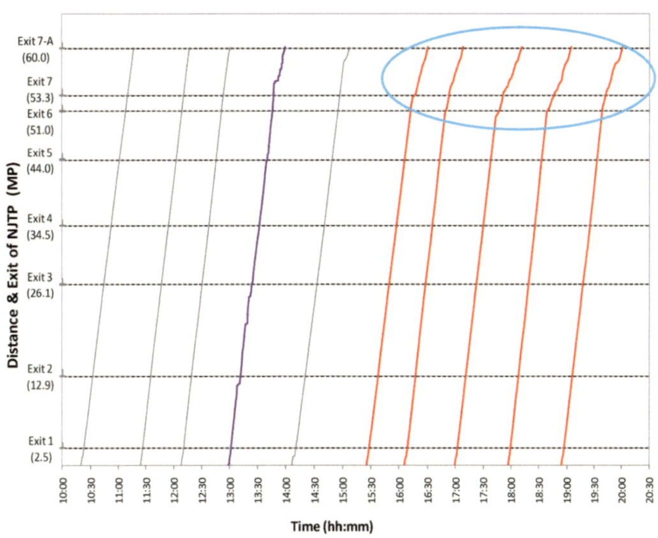

Fig. 3. Distance vs. Time on NJTP (A) (July 19, 2009).

departed earlier. Five vehicles, highlighted in red, experienced congestion beginning at Exit 6. The congestion occurred between Exit 7 and 7-A are the major cause for the high standard deviation of travel time/speed on July 19.

In addition to the Link-based analysis, the mean travel time and corresponding speed of NJTP and I-295 corridor are summarized in Table 5, and the findings are discussed below.

	Date	Mean (minutes)			STD (minutes)		
		NJTP (A)	NJTP (B)	I-295	NJTP (A)	NJTP (B)	I-295
Path Travel Time	5/24/09	56.3	53.2	59.0	1.2	1.2	1.8
	6/07/09	57.9	54.4	58.7	5.5	5.5	2.4
	7/19/09	64.2	60.8	58.1	5.3	5.0	4.1
	Date	**Mean (mph)**			**STD (mph)**		
		NJTP (A)	**NJTP (B)**	**I-295**	**NJTP (A)**	**NJTP (B)**	**I-295**
Path Speed	5/24/09	64.0	66.4	65.5	1.1	1.4	2.1
	6/07/09	62.3	65.0	65.8	5.3	5.9	2.8
	7/19/09	56.2	58.1	66.5	4.8	5.0	4.8

Table 5. Travel Time and Speed Comparison between NJTP and I-295.

May 24, 2009

The mean travel time of NJTP (A) was 56.3 minutes, which was shorter than that of the I-295 segment (59.0 minutes). The standard deviation of the mean travel time of NJTP (A) was 1.2 minutes, which was less than that of I-295 (1.8 minutes). Thus, the travel time of NJTP (A) segment was shorter and more reliable than that of the I-295 segment on 5/24/09. Travelers would be recommended to use NJTP instead of I-295 because of shorter and more reliable travel time.

June 7, 2009

The mean travel time on NJTP (A) was 57.9 minutes and the standard deviation of the mean was 5.5 minutes. On the other hand, the mean travel time on I-295 was 58.7 minutes, which was longer than that of NJTP (A); however, the standard deviation to the mean was 2.4 minutes, which was significantly less than that of NJTP (A). Thus, using NJTP (A) was not always quicker than I-295, and the reliability of travel time of the NJTP was less than that of I-295 on 6/7/09. Travelers would have been better using I-295 if they departed between 14:00 and 15:00, as they would have avoided congestion between Exits 7 and 7-A.

July 19, 2009

The mean travel time on NJTP (A) was 64.2 minutes and the standard deviation to the mean was 5.3 minutes. The mean travel time on I-295 was 58.1 minutes, shorter than that of NJTP (A), and the standard deviation to the mean was 4.1 minutes. It was found that using I-295 was quicker than NJTP (A) at the departure time of 13:00 through 19:00, and the reliability of travel time of I-295 was more than that of NJTP (A). Travelers would have been better using I-295 if they departed between 15:00 and 19:00, as they avoided congestion beginning at Exit 6.

Based on the mean speed and standard deviation, the travel time of these two parallel freeways were compared to identify the bottlenecks, evaluate their reliabilities and propose strategies to address congestion problems. For example, the observed congestion on NJTP (A) was found between Exits 6 and 7-A. Thus, it is recommended to use Dynamic Message Sign (DMS) at interchange 4 and Highway Advisory Radio (HAR) to inform drivers of congestion ahead.

4.2 Case study II – Analysis of adverse weather impact to travel time variability and reliability

The second set of travel time data intended for assessing weather impact were obtained through TRANSMIT system (TRANSCOM's System for Managing Incidents & Traffic). TRANSMIT uses vehicles equipped with electronic toll-collection tags (EZ-Pass) detected by roadside readers to approximate real-time travel time. The travel time data between 1/29/2008 and 2/29/2008 were collected on a 40-mile segment of the Interstate I-287 consisting of six links divided by seven TRANSMIT readers. The weather information was collected from a website (http://www.wunderground.com/US/NJ) managed by a weather data provider. Adverse weather conditions, such as rain, fog and snow, occasionally cause delay on roadways. The variation of travel times due to adverse weather should be concerned by agencies managing traffic operations and evaluating network-wide mobility. The measured variability and reliability indices discussed earlier will also help in trip planning under different weather conditions. The travel time data were collected by the TRANSMIT readers under three weather conditions, normal (dry), rain, and snow, during the AM peak period (6:00 AM ~ 9:00 AM) on weekdays. The mean and standard deviation of travel time, the 95th percentile travel time and buffer index were analyzed to quantify the impact of adverse weather to traffic conditions.

The day-to-day and within-day travel time were collected between 1/29/2008 and 2/29/2008. The start and end mile-posts (MP) and geo-coordinates of the TRANSMIT readers are summarized in Table 6. Link 6 has very long stretch, which is 20 miles from Exit 21 to Exit 41 on I-287. In Table 7, the link specific data are summarized, which includes link ID number and length. The mean, the 95th percentile, and the standard deviation of travel time as well as buffer index of each link are estimated by the equations discussed earlier in this paper.

Reader No	Reader ID	Reader Name and Location	Latitude	Longitude
A	5361664	NJ440 @ New Brunswick Av (MP 2.66)	40.521902	-74.291248
B	5361665	I-287 @ Route 1 (MP 0.93)	40.529952	-74.348758
C	5361666	I-287 @ Old New Brunswick Road (MP 7.71)	40.556262	-74.471249
D	5361667	I-287 @ Route 28 (MP 13.50)	40.565573	-74.553362
E	5361668	I-287 @ US202/6 (MP 17.66)	40.59809	-74.624353
F	5361669	I-287 Brunt Mills Rd (MP 21.44)	40.6462	-74.646085
G	5361671	I-287 @ I-80 (MP 42.20)	40.863936	-74.41704

Table 6. TRANSMIT Reader Information.

Link No	Link ID	Link Description and Location	Link Length (miles)	Free Flow Speed (mph)
1	5361672	NJ440 S MP 2.66 - I-287 N MP 0.93	3.59	75
2	5361673	I-287 N MP 0.93 - I-287 N MP 7.71	7.10	72
3	5361674	I-287 N MP 7.71 - I-287 N MP 13.50	5.79	71
4	5361675	I-287 N MP 13.50 - I-287 N MP 17.66	4.10	72
5	5361676	I-287 N MP 17.66 - I-287 N MP 21.44	3.84	73
6	5361691	I-287 N MP 21.44 - I-287 N MP 42.20	20.46	71

Table 7. TRANSMIT Link Information.

Day-to-day Travel Time Variation

Figure 4 indicates the mean and standard deviation of travel time on each link over different days. Results, in general, show that the mean travel times on Tuesdays and Fridays are relatively higher than other days. The greatest travel time happened on Link 6 which is not surprising because the link distance is much longer than others. The greatest standard deviation of travel time was observed on Link 3 on Fridays. It is worth noting that the length of Link 3 is 5.79 mile, and the mean travel time was 6 minutes, but the standard deviation of travel time was 5 minutes. The ratio of the standard deviation of travel time to the mean travel time of Link 3 on Fridays was extremely high.

The spatial and temporal distributions of the 95th percentile travel time of Link 6 are analyzed and shown in Figure 5, where the day-to-day travel time on each link seemed following a trend of mean travel time distributions shown in Figure 4. The motorists traveling on Tuesdays and Fridays generally suffered more congestion as well as the travel time uncertainty.

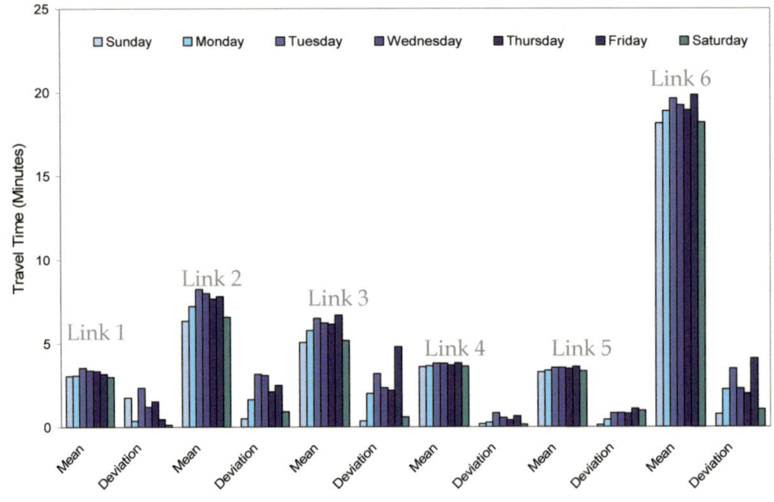

Fig. 4. Distributions of Mean and Standard Deviation of Travel Time.

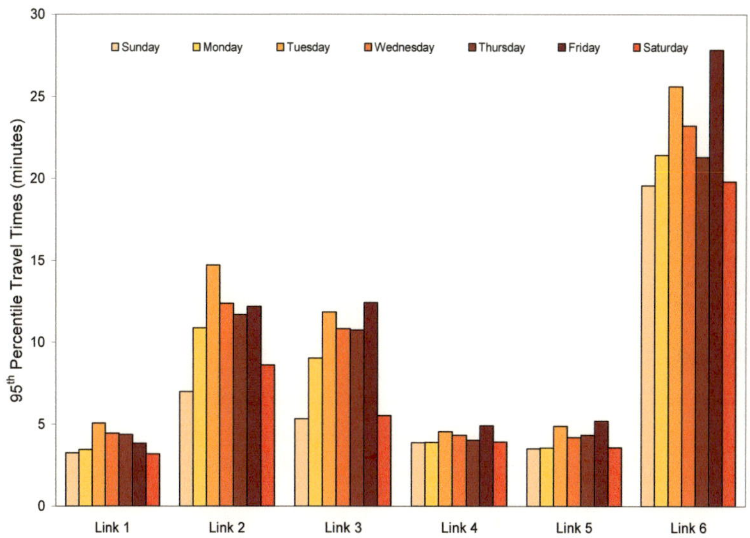

Fig. 5. Distributions of 95th Percentile Travel Time.

Within-day Travel Time Variation

The within-day variation of travel time is illustrated in Figure 6. It was found that the AM peak started before 7 AM and terminated around 9 AM and the PM peak started around 4 PM and terminated around 7 PM. The mean travel time of Link 6 is obviously higher than that of other links due to longer link length, which is consistent with the observations illustrated in Figures 4 and 5.

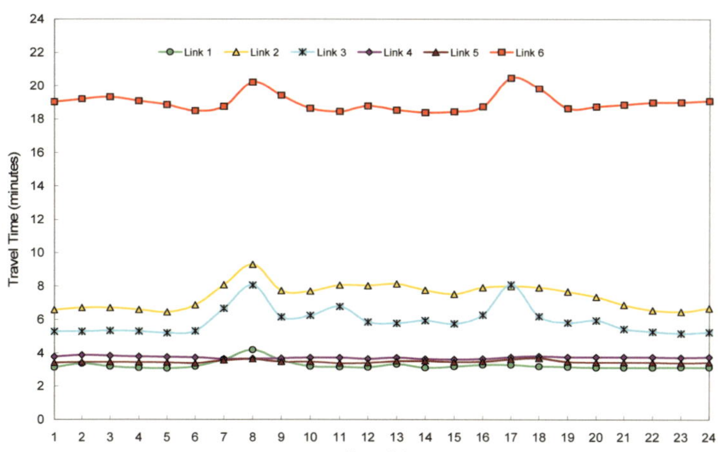

Fig. 6. Within-day Distributions of Link Travel Time.

Buffer Time and Buffer Index

As shown in Figure 7, the buffer time is the difference between the 95th percentile travel time and the mean travel time of Link 6. It was found that the two longest buffer times occurred in the AM and PM peak periods at 9 AM and 6 PM, respectively, while the shortest buffer time occurred during nighttime between 9 PM and 6 AM.

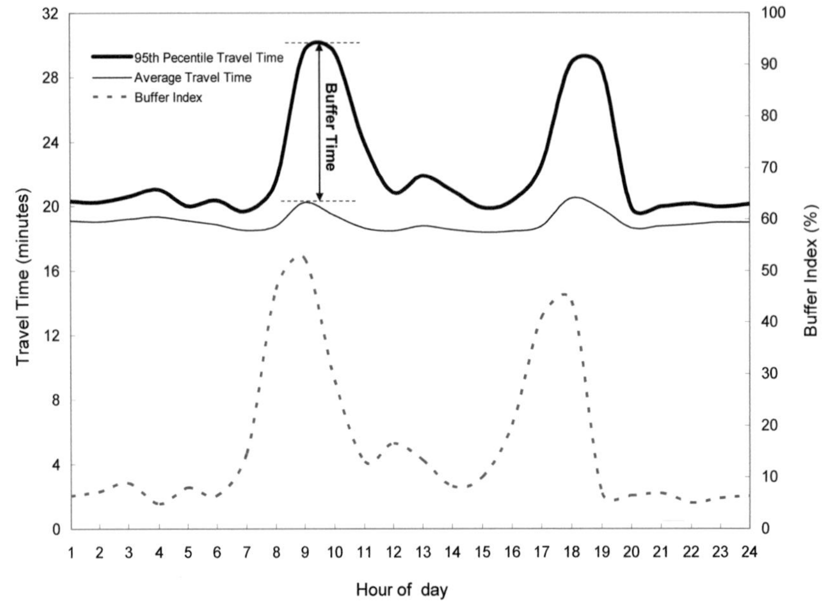

Fig. 7. Buffer Time and Index Distributions of Link 6.

Planning Time and Planning Time Index

The planning time, shown in Figure 8, is the difference between the 95th percentile travel time and free flow travel time of Link 6. Similar to buffer time, the greatest planning times of

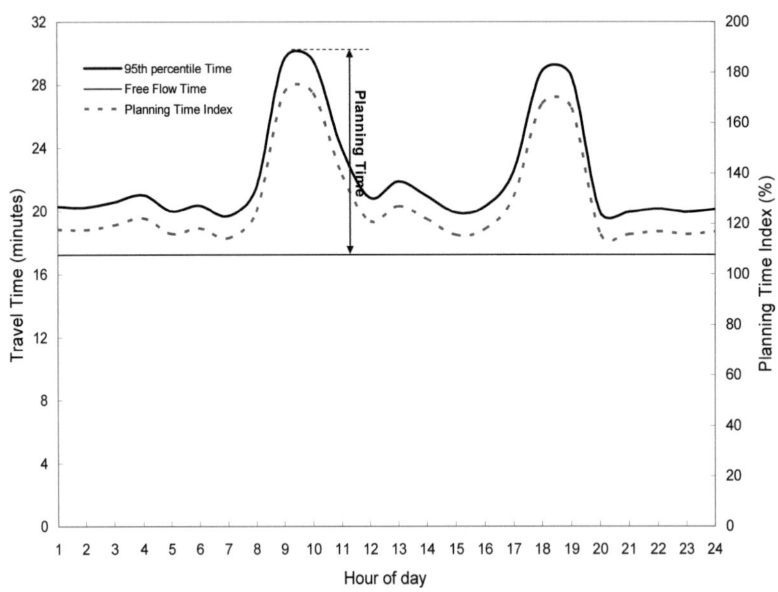

Fig. 8. Planning Time and Index Distributions of Link 6.

Link 6 occurred in the AM and PM peaks. As indicated in the above sections, the planning time index compares the worst case travel time to the travel time consumed under free-flow traffic condition.

Considering the whole studied I-287 segment, the travel time increased 10% and 60% under rain and snow, respectively. Moreover, it was found that the impact of adverse weather to traffic operation was much more significant during the peak periods than that during the off-peak periods.

The buffer Index of each link was calculated using Eq. 3 based on data collected under normal weather condition (18 days with dry pavement). To well plan for a journey, it is critical to know the buffer index, especially during the peak periods. In this study, the buffer indices of the AM and PM peaks were illustrated in Figure 9. In the AM peak, the worst buffer index was found on Link 1 with 75% while the best one was on Link 4 with 11%. In the PM peak, the worst buffer index was on Link 3 with 86%, while the best one was on Link 4 with 12%.

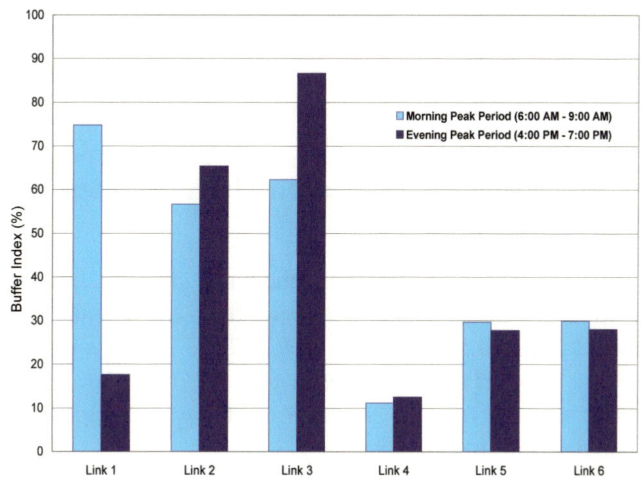

Fig. 9. Buffer Indices of Links in the AM and PM Peaks.

To ensure on-time arrival, the buffer time, may be estimated from the buffer index to account for unexpected delays. For example, given the mean travel time of Link 6 on Monday during the AM peak period is 18.9 minutes with a buffer index of 29%, the motorist should allow 5.5 minutes (=0.29*18.9) of extra time. Thus, the motorist should plan a total of 24.4 (=18.9+5.5) minutes to ensure on-time arrival at the end of Link 6.

Adverse weather is a critical factor contributes delay and travel time uncertainty due to reduced speed resulting insufficient road capacity. Due to limited data under adverse weather (e.g., snow and rain) occurred during the study period, the travel time data collected by the TRANSMIT readers under three weather conditions, such as normal (dry), rain, and snow, during the AM peak period (6:00 AM to 9:00 AM) on weekdays on Link 6 were analyzed. There were 18 dry-day, 4 rain-day, and 2 snow-day.

Considering adverse weather during the AM peak on Link 6, the collected data were classified by weather conditions, in which the mean and standard deviation of travel times, the 95th percentile travel time and buffer index were analyzed and illustrated in Figures 10. It was found that all travel time measures as well as the buffer index increase as the weather varying from dry to rain to snow condition.

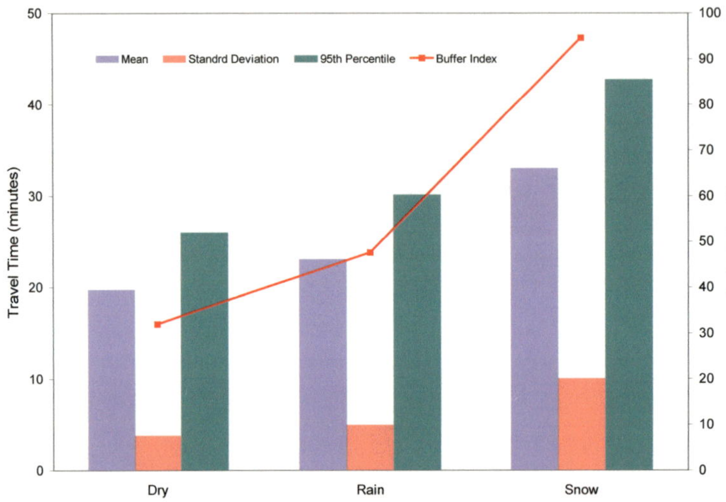

Fig. 10. Mean, Standard Deviation, the 95th Percentile Travel Time, and Buffer Index under Different Weather Conditions (AM Peak on Link 6).

Based on the buffer index shown in Figure 10, the variation (standard deviation) of travel time increased due to the weather condition varied from dry to rain and rain to snow. In addition, the buffer index also increased from 29%, 45%, to 94%. If a motorist needs to travel on Link 6 under rain, 16% (45%-29%) of extra time of the mean travel time under dry condition should be expected. However, under a snow day, the addition of 65% (94%-29%) of the mean travel time should be expected.

It was found that adverse weather, such as rain and snow, had significant impact on delay and associated travel time variability when compared with traffic operation under dry condition. Considering the whole studied I-287 segment, the travel time increased 10% and 60% under rain and snow, respectively. Moreover, it was found that the impact of adverse weather to traffic operation was much more significant during the peak periods than that during the off-peak periods.

4.3 Case study III – Investigating travel time variability in a freeway network

The third set of travel time data were used to analyze the travel time reliability under recurring and non-recurring traffic congestion. Travel time collection was conducted on the segments of fifteen New Jersey highways including: Routes US 1, NJ 4, US 9, NJ 17, US 22, NJ 24, NJ 29, NJ 42, US 46, NJ 70, NJ 73, I-76, I-78, I-80, I-280, and I-287 during the AM peak hours on weekdays between October 8, 2007 and April 21, 2008 through the use of Co-Pilot™ GPS navigation devices. In order to identify the link- and path-based travel time distributions of NJ 17, the following analysis was conducted:

The frequency of the route travel time for the entire departure time period from 6:15 to 8:15 AM is shown in Figure 11. The path travel time data were classified into four-minute time intervals, and the distribution is a shifted log-normal distribution. The Y- axis is intercepted at the free flow travel time then it is followed by a bell shape and then it has a long tail of path travel time observations that reflect the delays experienced due to various traffic flow conditions such as increases in demand and incidents.

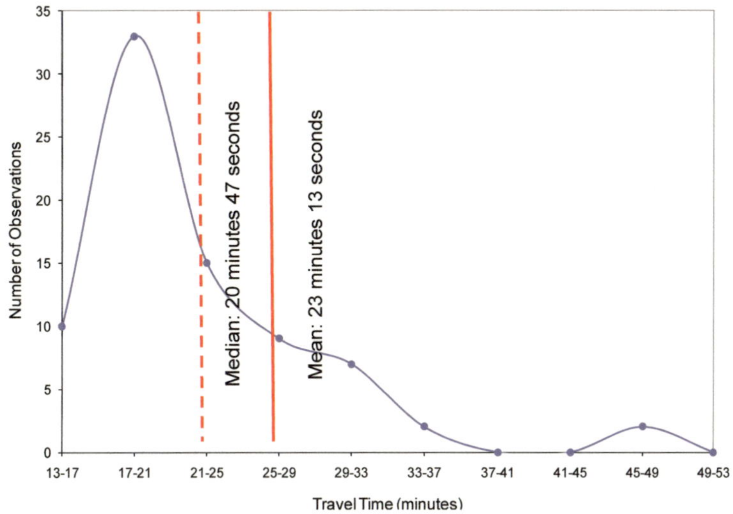

Fig. 11. Path Travel Time Distribution (NJ 17).

A normality test was conducted for NJ 17 and other routes that had a sufficient sample size. Normality tests are for testing whether the input data is normally distributed by some statistical tests such as Student's t-test, one-way and two-way ANOVA. In general, the normality test can be performed by using options such as Anderson-Darling test, Shapiro-Wilk test, and Kolmogorov-Smirnov test, which can be selected based on the sample size.

In order to determine whether a travel time distribution is normal or log-normal, hypothesis test was applied. The smaller the p-value, the more strongly the test rejects the null hypothesis. A p-value of 0.05 or less rejects the null hypothesis. The statistical statement for the normality test is

- Null Hypothesis(H_0): The route travel time distribution for NJ 17 follows the Normal distribution
- Alternative Hypothesis (H_a): The route travel time distribution for NJ 17 has a different distribution than the Normal.

Figure 12 indicates that the null hypothesis - the route travel time distribution is normal - is rejected as quite a few observations fall away from the line. The p-value (<0.005) of less than 0.05 confirms the visual observation of the normality test.

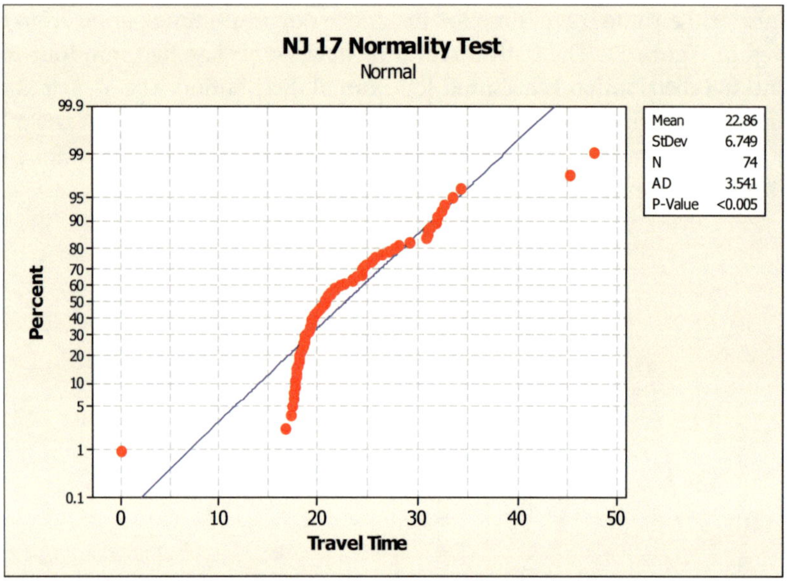

Fig. 12. Path Travel Time Distribution for Normality Test (NJ 17).

A Log-Normality test for the path travel time was also conducted. The collected path travel time was converted log-value and the shortest travel time was subtracted from each observation since the travel time distribution seems to fit a shifted log-normal distribution. Figure 13 indicates that the null hypothesis (e.g. the route travel time distribution is log-normal) is not rejected as only a few observations fall away from the line. The corresponding p-value (0.36) of greater than 0.05 confirms the visual observation of the log-normality test.

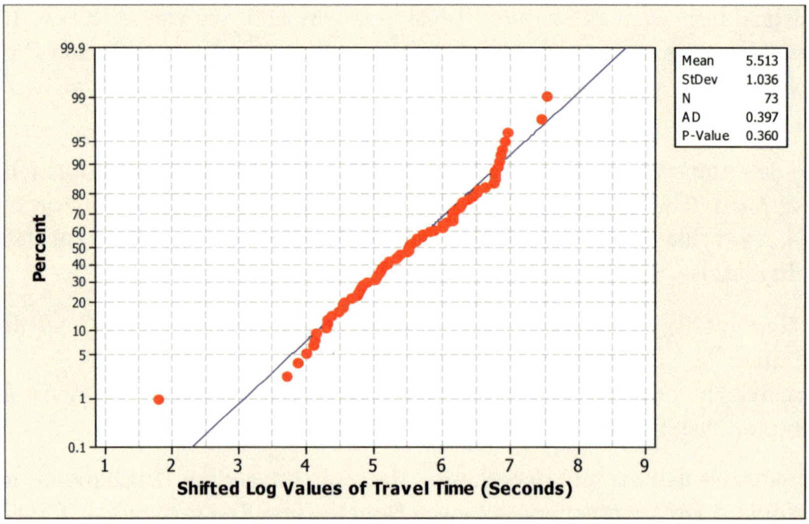

Fig. 13. Path Travel Time Distribution for Log-Normality Test (NJ 17).

The results of two tests are summarized in Table 8 for other routes.

Locations	Normality Test (p-value)	Log-Normality Test (p-value)	Log Mean	Log Standard Deviation	Log 95% C.I.
	-	√ (0.36)	5.5	1.06	0.22
NJ 208 / NJ 4	√ (0.083)	-	6.9	0.94	0.26
I-80 / I-280	-	√ (0.095)	5.5	1.04	0.25
NJ 24 / I-78	-	-	2.1	0.74	0.14
US 46 / NJ 3	-	√ (0.051)	5.7	0.85	0.19
US 22	-	√ (0.058)	5.9	1.00	0.18
I-287 (A)	-	√ (0.056)	5.7	0.85	0.18
I-287 (B)	-	-	5.3	0.90	0.18

√: p-value is greater than 0.05

Table 8. Normality and Log-Normality Test Results.

The majority of the path travel time distributions [e.g., NJ 17, I-80 / I-280, US 46 / NJ 3, US 22, I-287 (A), and I-287 (B)] were shown – using normality and log-normality tests - to follow a shifted log-normal distribution for the morning period from 6:00 to 9:00 AM. Specifically route NJ 208 / NJ 4 is the only one that shows some semblance of a Normal distribution. For each of these distributions the associated 95% Confidence Intervals were estimated. Due to limited data for each 15-minute time interval the corresponding path travel time distributions could not be estimated.

Based on the collected data, the corresponding path travel time distribution and the associated parameters –(e.g., mean, mean plus or minus standard deviation, median, and the 95th percentile of individual travel time, etc.) were calculated. The median of travel time is about 20.8 minutes, and the mean travel time is 23.6 minutes. The mean travel time is 46th of 73 samples, which is about 62% of the data below the mean. This indicates that the distribution has more samples on the low side of the mean and higher variation on the high side of the mean. The impact of departure time on path travel time is depicted in Figure 14. The mean path travel time with its deviation (e.g. plus and minus standard deviation) and the 95th percentile of individual travel time were estimated for each time period. Figure 14 shows that the travel time distribution for various departure time periods, the period of 8:30 - 9:00 in the AM peak had a significantly larger travel time variation, compared to that in the period of 6:30-7:00 AM. The delay observed in each time period was caused by construction activity on the roadside and increased demand at signalized intersections.

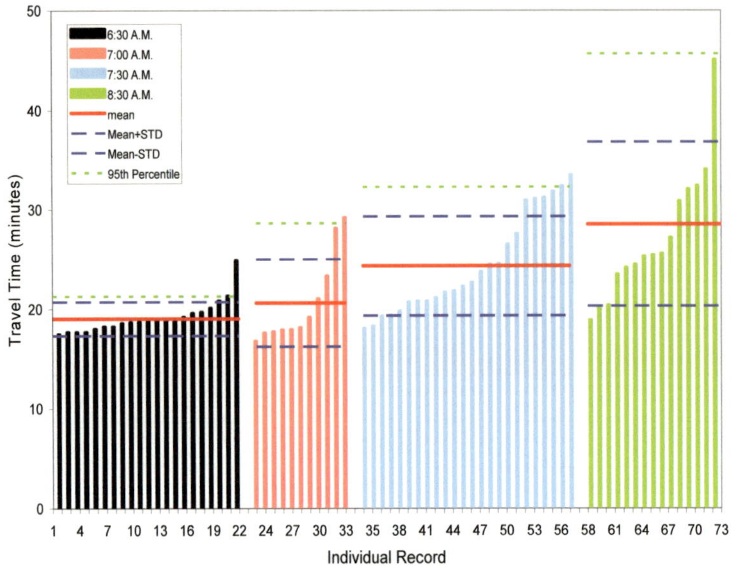

Fig. 14. Path Travel Time Distribution by Departure Time Period (NJ 17).

5. Conclusions

The chapter discussed methods to estimate the variability and reliability of travel times. Floating cars with GPS devices were proven as an applicable tool to collect travel time. The travel time variability and reliability were successfully assessed based on the mean travel time, the 95th percentile travel time, travel time index, buffer index, and planning time index, etc. The impact of stochastic factors on the corridor travel time was identified for the traffic management by public agencies. The GPS-based link/path travel time data could be used for producing the Measures Of Effectiveness (MOEs) thereby assisting real-time traffic control (e.g. signal timing) and traveler information (e.g. traffic diversion and travel planning), and transportation planning (e.g. infrastructure and traffic management strategies).

Many public transportation agencies have been focusing on enhancing the capability of data collection, processing, analysis and management to generate reliable estimates. In most cases, however, travel time data are available for relatively few corridors in a state and sometimes suffered by limited sample sizes. It is desirable to collect vast data in good quality and develop a cost-effective method to measure travel time variability and reliability that will be of interests of practitioners in measuring and predicting the transportation network performance. This study investigates the variability of travel time measurements based on broad traffic environments and their impacts on the ITS applications. It is recommended that the developed methodology can be used to produce link- and path-based travel time distributions associated with the Buffer Indices as performance measures to aid planners and engineers in making decisions of highway improvements and transportation management to reduce delays while improving reliability.

6. Acknowledgement

The authors would like to acknowledge the support of the New Jersey Department of Transportation for this study and Drs. Kyriacos Mouskos and Kitae Kim for data collection, processing and analysis.

7. References

Palma, A. & Picard, N. (2005). Route choice decision under travel time uncertainty, Transportation Research Part A Vol. 39 pg. 295–324.

Bartin, B. & Ozbay, K. (2006). Evaluation of Travel Time Variability in New Jersey Turnpike - A Case Study, Proceedings of the IEEE ITSE, Toronto, Canada.

Brownstone, D. & Small, K. (2005). Valuing time and reliability: assessing the evidence from road pricing demonstrations, Transportation Research Part A 39, pg. 279–293.

Smith, B., Zhang, H., Fontaine, M. & Green, M. (2003). Cell-phone probes as an ATMS tool, Final report of ITS Center project *for* University of Virginia Center for Transportation Studies. http://cts.virginia.edu/docs/UVACTS-15-5-79.pdf

Hellinga, B. & Fu, L. (2002). Reducing bias in probe-based arterial link travel time estimates, *Transportation Research* Part C vol. 10, no. 4, pp. 257-273.

Avineri, E., & Prashker, J. (2005). Sensitivity to travel time variability: Travelers' learning perspective, *Transportation Research Part C: Emerging Technologies,* Volume 13, Issue 2, pp. 157-183.

Lu, J., Yang, F., Ban, X. & Ran, B. (2005). Moments Analysis for Improving Decision Reliability Based on Travel Time, Transportation Research Record 1968, Jul. 2005, pp 109-116.

Sohn, K. & Kim, D. (2008). Dynamic Origin–Destination Flow Estimation Using Cellular Communication System , IEEE Transactions on Vehicular Technology - IEEE TRANS VEH TECHNOL , vol. 57, no. 5, pp. 2703-2713.

Chu, L., Oh, J. & Recker, W. (2005). Adaptive Kalman Filter Based Freeway Travel time Estimation, Transportation Research Board Annual Meeting, Washington D.C.

Elefteriadou, L. & Cui, X. Review of Definitions of Travel Time Reliability, In: Report for Florida Department of Transportation, Available from http://www.dot.state.fl.us/planning/statistics/mobilitymeasures/define-ttr.pdf

Lyman, K. & Bertini, R. (2008). Using Travel Time Reliability Measures to Improve Regional Transportation Planning and Operations, Journal of Transportation Research Board, TRR No. 2046, pp 1~10.

Fontaine, M. Yakala, A. & Smith, B. (2007). Final Contract Report: Probe sampling Strategies Traffic Monitoring Systems Based on Wireless Location Technology, Virginia Transportation Research Council.

SAFETEA-LU Intelligent Transportation Systems (ITS) Reporting and Evaluation Guidelines, Available from http://www.its.dot.gov/evaluation/eguide_safetealu.htm

Traffic Congestion and Reliability: Trends and Advanced Strategies for Congestion Mitigation, (2005). Federal Highway Administration Report, prepared by Cambridge Systematics, Inc. , Texas Transportation Institute.

Travel Time Data Collection Handbook, Report No. FHWA-PL-98-035

Yamamoto, T., Liu, K., & Morikawa, T., (2006). Variability of Travel Time Estimates using Probe Vehicle Data, Proceedings of ICTTS 2006, ASCE Vol.3, pp.278-287.

Recker, W. Chung, Y., Park, J., Wang, L., Chen, A., Ji, Z., Liu, H., Horrocks, M. & Oh J. (2005). Considering Risk-Taking Behavior in Travel Time Reliability, California PATH Research Report.

Sen, S. & Pillai, R. (2001). A Mean-Variance Model for Route Guidance in Advanced Traveler Information Systems, Transportation Science 35, pp. 37-49.

Zhou, Z. (2008). Models and Algorithms for Addressing Travel Time Variability: Applications from Optimal Path Finding and Traffic Equilibrium Problems. All Graduate Theses and Dissertations, Paper 129. Available from http://digitalcommons.usu.edu/etd/129

3

ITS Applications in Developing Countries: A Case Study of Bus Rapid Transit and Mobility Management Strategies in Dar es Salaam – Tanzania

Philemon Kazimil Mzee and Emmanuel Demzee
Transportation Management College
Dalian Maritime University, Dalian
PR China

1. Introduction

Mobility problem and planning technique suffered a revolution in the 1980s. We still encounter many of the same transport problems of the 1960s and 1970s: congestions, technology, pollution, accidents, financial deficits and so on. However, it was possible to learn a good deal from a long period of weak transport planning, limited investment, emphasis on the short term and mistrust in strategic transport modeling and decision making. A new contemporary dimension is the fact that most developing countries are suffering serious transport problems as well. These are no longer just the lack of roads to connect distant rural areas with markets. Indeed, the new transport problems bear some similarities with those prevalent in the industrialized world: congestion, pollution, technology, and so on. Fortunately, transportation technologies and mobility management strategies are emerging that can help to meet the transportation challenge; these include ITS technology that provide and foster more reliable operation guidance, information, fare technology, automotive, fuel technologies and so on. For the purpose of this chapter we assume that intelligent transport system (ITS) should be taken to include all those systems that use information technology. The impact of information system and other ITS innovation on travel demand and transport system performance can be modeled in many different ways, but the key ways of the modeler is to determine the appropriate level of aggregation. This will be determined primarily by the purpose of the exercise. At one extreme strategy planner might welcome an aggregate model which uses information theory to predict the consequences that ITS innovation should have for the efficiency of the transport system. Implementation of Bus Rapid Transit - BRT and mobility management strategies can enhance the mobility in the city and reduce the demand for private vehicles. It can be concluded from this chapter that the introduction of a BRT system in Dar es Salaam has numerous evident gains in terms of improving the current public transport supply. Nevertheless, some shortcomings have been identified that limit the contribution of the system in its endeavor towards creating a sustainable urban mobility system in the city. These shortcomings are the inequitable distribution of services across population groups; the need to limit the growth in the number of cars; the lack of involvement of the current public transport operators in the planning process and the heedless handling of their fates. Mobility and transportation are also the leading indicators of economic growth of a society. Unfortunately, if left unchecked, these indicators show a declining trend with the passage of time (i.e., traffic congestion) because transportation systems are often designed to overcome the present crisis without considering the increasing nature of the population of a country. Under Dar es Salaam city in Tanzania the conditions of accessibility of existing public transport and mobility is deeply creates great hardship for the citizen they have bad seat arrangements, overload passengers, not follow allocated routes, speed and drive recklessly, emit large amount of large amount of pollutants and Carbon Dioxide (CO2), and so on. Hence you can find that in developing countries can quickly reveal the source of customer dissatisfaction with public transport and non-motorized options poor transit services and these push consumers to private vehicle options. The splendid transit system called Bus Rapid Transit System with more mobility when compare to heavy rail/ light rail transit. It is an integrated, well defined system with design features similar to light rail rapid transit systems. BRT represents a way to improve mobility at a relatively low cost through incremental investment in a combination of bus infrastructure, operational movements and technology. BRT will utilize intelligent transportation system technology, modern land use planning and transportation policies to support new concept for rapid transit system.

2. Over view of population and urbanization

2.1 Population growth and urbanization

Although travel has been part of a human experience for many centuries, the 20th century seems to have depicted a quite distinct experience (OECD, 2000). The century happens to be characterized by remarkable rates of growth in population, mobility and urbanization such that, globally, motorized transport increased by more than one hundred fold while population increased by fourfold (OECD, 2000). Along with these changes, urban land has been rapidly expanding due to the modern patterns of city growth that are land intensive as well as the improvements in transportation services that made commuting easier (UNFPA, 2007). The sum total of these phenomena has thus been a continuing movement of people resulting in tremendous growth of urban areas (UNFPA, 2004). Accordingly, at present, for the first time in history, more than half of the world's population dwells in cities and it is estimated that all regions of the world will have urban majorities by 2030 (Habitat, 1996a). This gives urban areas, and the issues related to them, an increasing importance in contemporary socio - economic and environmental discussions.

2.2 Population trends in Tanzania

The country has experienced continued, steady population growth over the past three decades. According to the data from National Population and Housing Census, which has been conducted four times by the National Bureau of Statistic (NBS), the population of Tanzania mainland has increased nearly triple since 1967; the population extended from 11.9 million persons in 1967 to 17.0 million persons in 1978, 22.4 million persons in 1988 and to 33.4 million persons in 2002 (Table 2.1). The average annual population growth rate was 3.3 percent between 1967 and 1978, 2.8 percent between 1978 and 1988 and 2.9 percent between 1988 and 2002. Tanzania represents one of the larger African countries on a population basis. According to the World Bank population index of year 2004, Tanzania had fifth largest population size following Nigeria, Egypt, Ethiopia and South Africa. Population size is roughly comparable to Sudan and Kenya (Table 2.2). Population density is relatively modest at approximately 43 persons per square kilometer of land area. The share of urban populations was 36.4 percent in Tanzania, which was lower than South Africa (57.4 percent), Kenya (40.5 percent) and Sudan (39.9 percent) and much higher than Ethiopia (15.9 percent) and Uganda (12.3 percent).

Year	Tanzania Mainland	Average Annual Growth Rate (percent per annum)
1967	11,958,654	
1978	17,036,499	3.27 (1 967-1 978)
1988	22,455,207	2.80 (1 978-1 988)
2002	33,461,849	2.89 (1988-2002)

Source: The United Republic Tanzania Population and Housing Census, NBS.

Table 2.1. Population Trends in Tanzania Mainland.

2.3 Population growth and urbanization in Dar es Salaam City

The increasingly trend of urbanization worldwide can be demonstrated using the rapid increase over the past decades in urban population in almost all regions of the world (UNFPA, 2004). According to the United Nations World Urbanization Prospects (WUP, 2005), urban population has grown from 29% in 1950 and 37.2% in 1975 to 48.7% in 2005. That represented an annual rate of change of 2.65% which was more than double the rate for rural population, 1.12% (Habitat, 1996a). Following this increasing trend, it is estimated that the global urban population will reach as much as 5 billion by 2030, which would represent 60% of the world's population (UNFPA, 2004). Although both developed and developing countries have been experiencing an increase in urban population, the rates of change in these regions vary greatly. According to (WUP, 2005), the urban population in developing countries increased annually at 3.61% between 1950 and 2005 while developed countries had the increase at only 1.37% per year. Likewise, these rates are estimated to be 2.20% and 0.47% for developing and developed countries, respectively, over the years between 2005 and 2030 (Habitat, 1996a). This is quite plausible as internal migration (i.e. the movement of people within a country) from rural areas to urban areas is high in the developing world for several reasons (UNFPA, 2004). The most important ones appear to be the so called demand pull and supply push factors that denote productivity and monetary reward prospects in urban areas and rural poverty due to traditional agriculture. (Borris. S.L Bertinell & E. Strobl 2006) While social changes induced by need for higher education, rising incomes, and emerging life styles are other reasons (Button, K & P. Nijkamp 1997) Being an African city, Dar es Salaam has for decades been experiencing rapid population growth (Olvera, L.D Plat & Pochet 2003). States, for example that the population of the city grew at an annual growth rate of 9.4% between 1968 and 1978 and at 4.7% between 1978 and 1988. The past decades showed similar patterns such that the population of the city reached 2.5 million by 2002, doubling the population since 1988 (Lupala J 2002). Currently, the population growth rate of the city is estimated to be 4.1% (JICA, 2008). This denotes a pressure directly exerted on the transport demand in the city, be it in terms of the need for more vehicles or the need to commute longer due to a geographical growth of the city. Figure 2.1 below shows the map of Dar es Salaam city.

Country	Year 2004 Population	Land Area (sq km)	Population Density (Persons/sq km)	Percent Urban Population
Nigeria	128,710,000	910,770	141.3	47.5
Egypt	72,642,000	995,450	73.0	42.2
Ethiopia	69,961,000	1,000,000	70.0	15.9
South Africa	45,509,000	1,214,500	37.5	57.4
Tanzania	**37,627,000**	**883,590**	**42.6**	**36.4**
Sudan	35,523,000	2,376,000	15.0	39.9
Kenya	33,467,000	569,140	58.8	40.5
Algeria	32,358,000	2,381,700	13.6	59.4
Morocco	29,824,000	446,300	66.8	58.1
Uganda	27,821,000	197,100	141.2	12.3
Ghana	21,664,000	227,540	95.2	45.8
Mozambique	19,424,000	784,090	24.8	36.8
Madagascar	18,113,000	581,540	31.1	26.8
Ivory Coast	17,872,000	318,000	56.2	45.4
Cameroon	16,038,000	465,400	34.5	52.1
Angola	15,490,000	1,246,700	12.4	36.4
Niger	13,499,000	1,266,700	10.7	22.7
Mali	13,124,000	1,220,200	10.8	33.0
Zimbabwe	12,936,000	386,850	33.4	35.4
Burkina Faso	12,822,000	273,600	46.9	18.2
Malawi	12,608,000	94,080	134.0	16.7
Zambia	11,479,000	743,390	15.4	36.2
Senegal	11,386,000	192,530	59.1	50.3
Tunisia	9,932,400	155,360	63.9	64.1
Chad	9,447,900	1,259,200	7.5	25.4
Guinea	9,201,800	245,720	37.4	35.7
Rwanda	8,882,400	24,670	360.0	20.1
Benin	8,177,200	110,620	73.9	45.3
Somalia	7,964,400	627,340	12.7	35.4
Burundi	7,281,800	25,680	283.6	10.3
Togo	5,988,400	54,390	110.1	35.7
Libya	5,740,100	1,759,500	3.3	86.6
Sierra Leone	5,336,400	71,620	74.5	39.5
Eritrea	4,231,500	101,000	41.9	20.4
Central African Rep.	3,986,000	622,980	6.4	43.2
Congo, Rep.	3,882,900	341,500	11.4	53.9
Liberia	3,240,600	96,320	33.6	47.3
Mauritania	2,980,400	1,025,200	2.9	63.0
Namibia	2,009,300	823,290	2.4	33.0
Lesotho	1,798,000	30,350	59.2	18.1
Botswana	1,769,100	566,730	3.1	52.0
G. Bissau	1,539,700	28,120	54.8	34.8
The Gambia	1,477,700	10,000	147.8	26.1
Gabon	1,362,300	257,670	5.3	84.4
Mauritius	1,234,200	2,030	608.0	43.6
Swaziland	1,119,800	17,200	65.1	23.7
Djibouti	779,100	23,180	33.6	84.1
Comoros	587,940	2,230	263.7	35.6
Cape Verde	495,170	4,030	122.9	56.7
Eq. Guinea	492,230	28,050	17.5	49.0
Sao Tome & Principe	152,960	960	159.3	37.9
Seychelles	83,643	460	181.8	50.1

Source: Study Team based on World Bank data. Ranking in descending population size

Table 2.2. African Population Patterns in Year 2004.

2.4 Urban sprawl

In accordance with the observed population growth in urban centers, almost all cities throughout the world experience expansion in their geographical space (UNFPA, 2007). This assumes different reasons in different parts of the world. In the developed world, and particularly in North America, urban sprawl results largely as people move into suburb areas in search of a higher quality of living (White hand J. & J. Larkham, 1992). While in the developing world, and typically in African countries, urban sprawl occurs as people build illegal houses in the peripheries of cities and render cheap rental conditions thereby attracting more and more settlers (Habitat, 19996a). Driven largely by many factors, the surface area of Dar es Salaam has increased rapidly over the past decades. (Olvera, L. D. Plat & P. Pochet, 2003) State, for example, that the land coverage of the city increased by many factors between 1968 and 1982. Likewise (Lupala. J 2002) records the geographic growth of the city during the period between 1963 and 2001 to be an increase by more than 18 times (from 3,081 ha to 57, 211 ha). Presenting this differently, (Olvera, L. D. Plat & P. Pochet, 2003) shows that the distance from the city center to the edge has increased from 15 km in 1978 to 30 km by the mid 1990s. Currently Dar es Salaam has a size of about 1800 km/sq (Hall. F, 2004 & JICA 2008)

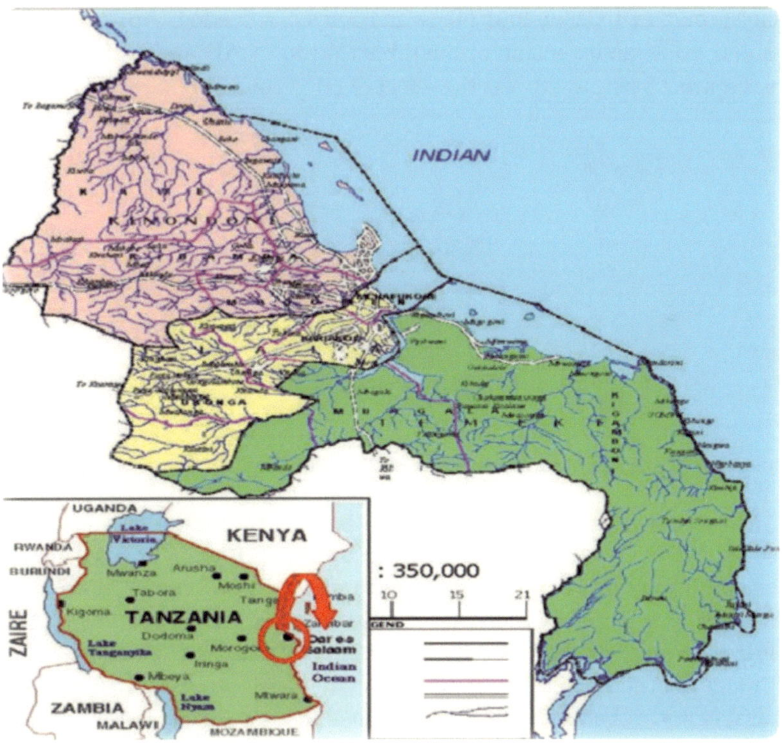

Fig. 2.1. Map of Dar es Salaam.

2.5 Designing model approach for the bus networks to grow with cities

This simple model looks at the rapid growth of a hypothetical city. The city is divided into zones of 1km square, each with a density of 200 inhabitants per ha, where bus passengers can reach a route by walking a maximum of about 500m. For the first stage, of 80,000, one simple "square" bus route, as marked in blue, can connect all the zones. If a new zone is added, this route can simply be extended from A to allow access to the entire city. Figure 2.2 below express the effect of city grow on bus routes (80,000 to 180,000)

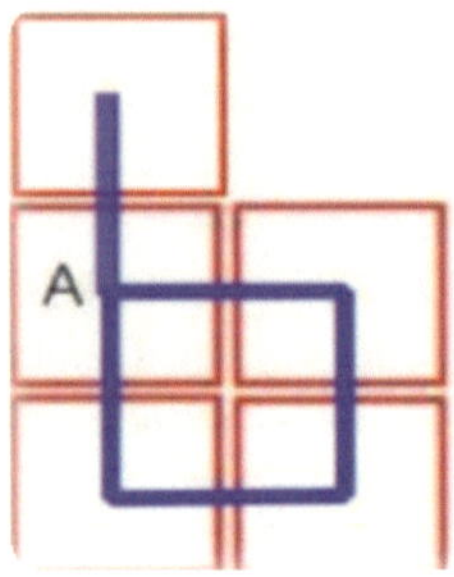

Fig. 2.2. The effects of city grow on bus routes (80,000 to 180,000).

When the city grows to 180,000, four (4) basic routes are needed: ABD; ACD; AD; and BC. Again, for a new zone, an extension of the three (3) routes AD and ABD and ACD give full accessibility. Figure 2.3 below express the effect of city grow on bus routes (180,000

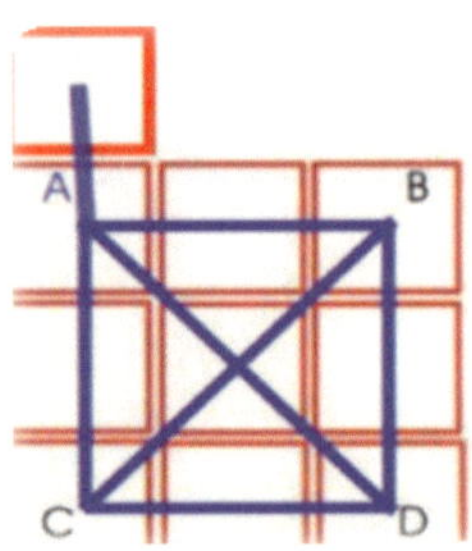

Fig. 2.3. The effect of city growth on bus routes 180,000.

In figure 2. 4: The city reaches a population of 320,000 by expanding in area, adding an extra 25% on each side. The route network needed to link the different zones starts to adopt the zigzag pattern typical of larger towns, and the routes have become longer 5 to 6 km instead of 2 to 4 km. eight routes are now needed for all zones to be connected: AMP; ADP; AP; MD; BFGL; CGFJI; EFJKO; HGKJN. There are also several routes superimposed, with excess capacity although this still leaves no direct connection between B and J; C and K; H and F; L and J; O and G; N and F; I and K; and E and G. Passengers between these zones must change buses and pay an extra fare. A new zone added to this structure will require six (6) extended routes, causing these routes to share the available demand and become 1km longer. This expansion is doomed to be unprofitable. Other typical problems of this "organic growth" are also shown: the purple route develops an irrational pathway; the black route has two (2) variations (say "Black a" and "Black b"), which means that the frequencies at the end points are unsatisfactory; and there is an oversupply of capacity at the route endings. If the city plans to develop new districts, with five (5) new zones at specific locations, each zone will need several new routes to reach the existing city areas, including a route that offers a direct link between the zones themselves. Figure 2.4 routes on grid for a city of 320,000

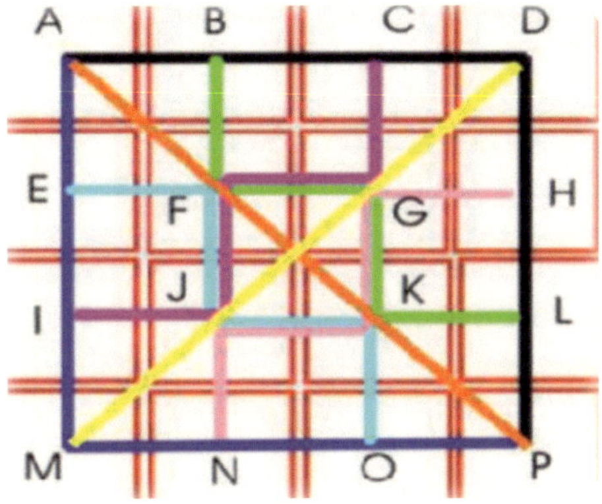

Fig. 2.4. Routes on a grid for a city of 320,000.

In the larger city an expanded system for a city of 320,000; needed each new development zone full accessibility to the main destinations (see Figure 2. 5). If this is not available, the passenger either has to use another mode to reach a satisfactory route, take more than one route, or simply not use public transport. At this stage in a city's development, there is a need to follow the integrated approach. This allows the system to expand in order to meet the demands of planned city growth. Figure 2.6 shows how expansion in two (2) main corridors, north - west and south - east, can be included into the existing system by adding two (2) new interchange terminals at C and D. The main trunk route becomes C to D, using high capacity vehicles and some form of bus priority. The extra ten (10) zones at the same density would have an additional population of 200,000, taking the city to 520,000 inhabitants. Once the basic infrastructure is in place, access to the entire transport network is simplified and thus can become an essential factor in promoting the desired urban development. Figure 2.5 and 2.6 below express system for a city of 320,000 and expended integrated system on a grid for a city of 520,000 respectively.

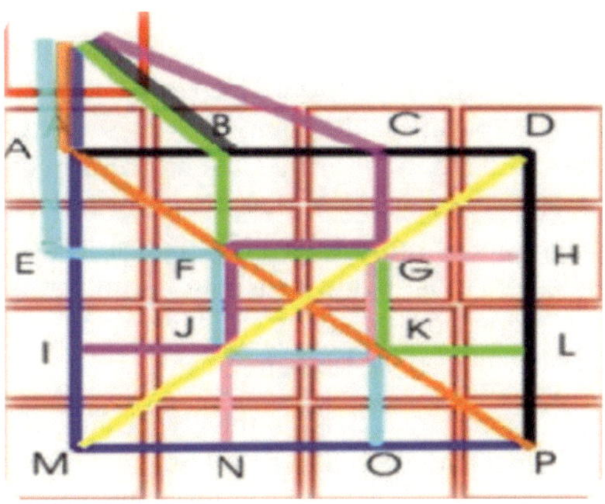

Fig. 2.5. An expanded system for a city of 320,000.

Fig. 2.6. Integrated system on a grid for a city of 520,000.

3. Mobility management and strategies in developing countries

3.1 Structural principles for sustainability

To achieve viability and sustainability of the transport systems in the developing countries, a set of structural principles are prepared:

i. A strategic policy must be set which integrates and coordinates all aspects of urban development and planning;
ii. The system needs to define single responsibility and accountability;
iii. Risk and responsibility must be appropriately assigned between parties (government and agency); No party should carry risk it cannot manage;
iv. The responsible agency needs adequate resources (self generating revenue mechanisms or allocated budgets) to perform its task;
v. The institution needs to be involved in setting its own internal business strategy in accordance with strategic policy guidelines;
vi. The necessity of a sound business model that delivers efficiency, sustainability, value for money (and integrated value added benefits), prioritization, economic, social and environmental benefits;
vii. Use of performance/outcomes - based contract to specify and document the undertakings of both parties and establish the working relationship between them. Normal contractual conditions will specify general terms and conditions while the specifications of service outlines the quantity of service, payments as well as performance indicators, and the monitoring and enforcement conditions
viii. The contract is commercially oriented in that it is fully funded to provide its services. No public service obligations or implications cloud its task and responsibilities;
ix. Key Performance indicators must relate clearly to outputs and outcomes (deliverables, measurable data) as evidenced in service delivery, or outcomes (quality of life indicators) showing policy objectives are being met.

For the case of Dar es Salaam city in Tanzania, the establishment of Dar es Salaam Rapid Transit Agency - (DART Agency) addressed in detail many of above principles and outlines in considerable detail the objectives and methodologies of implementation and administration of the agency. At a higher level, Dar es Salaam Urban Transport Authority

(DUTA) will provide direction and coherence across all sectors so they operate harmoniously. The Institutional Development in Dar es Salaam Transport Authority has proposed to establish the Dar es Salaam Urban Transport Authority (DUTA) as an accountable and transparent authority responsible for the transport system development of Dar es Salaam. The management of urban transport as a function is almost nonexistent in Dar es Salaam specifically due to the lack of guidance attention for responsible agencies and the number of responsible entities involved, with municipalities and also national authorities cross - cutting into urban transport affairs. This has resulted in a fragmented planning process and lack of coordination, vertically and horizontally between levels of government and departmental disciplines. As one of the strategic policies to meeting these challenges DUTA has been proposed and its conceptual design has to be presented.

3.2 Transport demand

Transport problems have become more widespread and severe than ever that in both industrialized and developing countries alike. Fuel shortage are (temporarily) not a problem but general increase in road traffic and transport demand has resulted in congestion, delay, accidents and environmental problems well beyond what has been considered acceptable so far. These problems have not been restricted to road and car traffic alone. Economic growth seems to have generated levels of demand exceeding the capacity of most transport facilities. The demand for transport service is highly qualitative and differentiated. There is a whole range of specific demands for transport which are differentiated by time of day, day of week, journey purpose, type of cargo, importance of speed and frequency, and so on. A transport service without the attribute matching this differentiated demand may well be useless. The demand for transport is derived; it is not an end in itself. With the possible exception of sightseeing, people travel in order to satisfy a need (work, leisure, education, health) at their destination. This is even more - true of goods movement. In order to understand the demand for transport, we must understand the way in which facilities to satisfy these human or industrial needs are distributed over space, in both urban and regional content. A good transport system widens the opportunity to satisfy these needs; a heavily or poorly connected system restricts options and limits economic and social development.

3.3 Transportation demand in Dar es Salaam City

Burdened with a rapid population growth and city expansion, Dar es Salaam's transport sector depicts a situation where the gap between public transportation needs and provision is continuously widening (Olvera *et al.*, 2003). This situation has been worsened as public transport is the only alternative for the poor. Two World Bank surveys done in 1994 and 1996 show, for example, that about 43% of all the trips done in Dar es Salaam were done using public transport, whereas only 7% of these trips were done with cars and motorbikes and 3% by bicycles (SSATP, 2005). A recent study by (Kombe *et al.* 2003) & (Kanyama *et al.*, 2004) also shows that as much as 60% of all trips made in the city are done on buses. Furthermore, the city's transport sector is burdened with high travel demand soaring from the uneven distribution of public and private facilities within the city (SSATP, 2005). This denotes, firstly, a concentration of employment and market opportunities in the city center. According to (Olvera *et al.* 2003), for example, Dar es Salaam's main urban facilities; its port, the main hospital, the largest market and the commercial district Kariakoo are all located in the center of the city which obliges the inhabitants to commute to access the opportunities there. And secondly, it denotes lack of services, as schools and health units, within or close to residential areas that makes commuting unavoidable. (Olvera *et al.* 2003) demonstrates this by analyzing a 1993 Dar es Salaam Human Resources and Development Survey that the closest public secondary school and the closest public hospital are located on average at 4.8 km and 3.9 km, respectively, away from any home in the city, with higher figures for poor neighborhoods. The private alternatives are on average 3.9 km and 5.4 km away (Ibid). Although the figures given above duly portray the level of stress on the public transport in the city, accurate information can be gained by looking at the overall travel demand. A 2007 survey done by the Japan International Cooperation Agency - JICA has established that 2.87 million trips per day are made in total in the city using all modes of transport and that 2.13 million (74%) of those trips are made by motorized modes (JICA, 2008).

3.4 Transport delivery

Over view of the characteristic of transport supply is that it is a service and not a good. Therefore, it is not possible to stock it, for example, to use it in times of higher demand. A transport service must be consumed when and where it is produced, otherwise its benefit is lost. For this reason it is very important to estimate demand with as much accuracy as possible in order to save resources by tailoring the supply of transport service to it. Many of the characteristics of transport system derive from their nature as a service. In very broad terms a transport system requires a number of fixed assets, the infrastructure, and a number of mobile units, the vehicles. It is the combination of these, together with a set of rules for their operation. That makes possible the movement of people and goods. It is often the case that infrastructure and vehicles are not owned nor operated by the same group or company. This is certainly the case of most transport modes, with the notable exception of many rail systems. This separation between supplier of infrastructure and provider of the final transport service generates a rather complex set of interaction between government authorities (central or local), construction companies, developers, transport operators, travelers and shippers and the general public. The latter plays several roles in the supply of transport service: it is usually the resident affected by a new scheme, or the unemployed in an area seeking improved accessibility to foster economic growth.

3.4.1 Interactive fuzzy mult-objective linear programming to transportation planning decision

In real-world transportation planning decision (TPD) problems, input data or related parameters are frequently imprecise / fuzzy owing to incomplete or unobtainable information, a novel interactive fuzzy multi objective linear programming (*i*-FMOLP) model for solving TPD problems with multiple fuzzy objectives. The proposed *i*-FMOLP model attempts to minimize simultaneously the total production and transportation costs and the total delivery time with reference to available capacities at each source and forecast demand at each destination. The transportation planning decision (TPD) problem involves the distribution of goods and services from a set of sources (e.g. factories) to a set of destinations (e.g. warehouses). With a variety of transporting routes and differing transportation costs for the routes, the aim is to determine how many units should be shipped from each source to each destination so that all demands are satisfied with the minimum total transportation costs. Basically, the TPD problem is a special type of a linear programming (LP) problem that can be solved using the standard simplex method. Some special solution algorithms, such as the stepping stone method and the modified distribution method, allow TPD problems to be solved much more easily than the general LP method. However, when any of the LP method or the existing effective algorithms is used to solve the TPD problems, the goal and related inputs are generally assumed to be deterministic/crisp (L .Li & K. Lai, 2000). In most practical TPD problems, input data or related parameters are often imprecise/ fuzzy owing to incomplete or unobtainable information. Obviously, conventional LP method and solution algorithms cannot solve all fuzzy TPD problems. In 1976, (Zimmermann, 1997) first introduced fuzzy set theory into an ordinary LP problem with fuzzy goal and constraints. Following the fuzzy decision-making method proposed by (Bellman & Zadeh, 1970), that the chapter confirmed the existence of an equivalent LP problem. Since then, fuzzy linear programming (FLP) has developed into several fuzzy optimization methods for solving TPD problems. (Chanas et al. 1984) presented an FLP model for solving the TPD problem with crisp cost coefficients and fuzzy supply and demand values. Moreover, (Chanas & Kuchta, 1998) proposed the concept of the optimal solution of the TPD problem with fuzzy coefficients expressed as L-L fuzzy numbers, and developed an algorithm for obtaining the solution. Related works on the use of FLP to solve TPD problems included (Bit et al. 2001, Chanas et al. 1984, and Chanas and Kuchta 1998). However, in real-world TPD problems, the decision maker (DM) must simultaneously handle conflicting aims that govern the use of the resources within organizations. These aims are minimizing total production costs, total transportation costs and total delivery time/distance, and maximizing total profits, total relative safety, customer service level and utilization of equipment and facilities (Bit et al. 2001, Chanas et al. 1984, and Chanas and Kuchta 1998). Particularly, the DM must simultaneously optimize these conflicting objectives in a framework of fuzzy aspiration levels. In 1978, (Zimmermann 1997) first extended his FLP approach to a conventional multi-objective linear programming (MOLP) problem. For each of the objective functions in this problem, the DM was assumed to have a fuzzy objective, such as "the objective function should be substantially less than or equal to some value". Subsequent works on fuzzy goals programming (FGP) included (Luhandjula 1982, Sakawa 1988, Chen and Tsai 2001). Subsequently, researchers have developed several FOP methods to solve multi-objective TPD problems. Bit et al. proposed an additive fuzzy programming model that considered weights and priorities for all non equivalent objectives for the multi objective TPD problem. (Li and Lai 2000) designed a fuzzy compromise programming method to obtain a non-dominated compromise solution for multi-objective TPD problems in which various objectives were synthetically considered with the marginal evaluation for individual objectives and the global evaluation for all objectives. (El-Wahed, 2001) developed a fuzzy programming approach to determine the optimal compromise solution of a multi objective TPD problem by measuring the degree of closeness of the compromise solution to the ideal solution using a family of distance functions. Related works on fuzzy multi objective TPD programming problems included (Bit et al. 2001). This thesis develops a novel interactive fuzzy multi objective linear programming (*i*-FMOLP) model for solving TPD problems with multiple fuzzy objectives. The proposed *i*-FMOLP model attempts to simultaneously minimize the total production and transportation costs and the total delivery time with reference to available capacities at each source and forecast demand at each destination.

3.4.2 Problem description, assumptions and notation

The TPD problem examined herein can be described as follows. Assume that a distribution center seeks to determine the transportation plan of a homogeneous commodity from m sources to n destinations. Each source has an available supply of the commodity to distribute to various destinations, and each destination has a forecast demand of the commodity to be received from the sources. The TPD proposed herein attempts to determine the optimal volumes to be transported from each source to each destination to simultaneously minimize the total production and transportation costs and the total delivery time. The TPD problem proposed in this work focuses on developing an interactive i-FMOLP model for optimizing the transportation plan in fuzzy environments. The mathematical model developed herein is based on the following assumptions.

i. All of the objective functions are fuzzy with imprecise aspiration levels.
ii. All of the objective functions and constraints are linear equations.
iii. The values of all model parameters are certain over the planning horizon.
iv. The transportation costs and delivery time on a given route is directly proportional to the units shipped.
v. The total supply available at all sources just equals the total demand required at all destinations.
vi. The linear membership functions are assigned to represent the fuzzy sets involved, and the minimum operator is used to aggregate all fuzzy sets.

Assumption (i) concerns the fuzziness of the objective functions in practical TPD problems, and incorporates the variations in the DM's judgments for the solutions of fuzzy multi objective optimization problems in a framework of fuzzy aspiration levels. Assumptions (ii) to (iv) indicate that the linearity, certainty and proportionality properties must be technically satisfied as a standard LP problem. Assumption (v) is the 'necessary and sufficient' condition for a feasible solution to the TPD problem. Assumption (vi) is made to convert the original fuzzy MOLP problem into an equivalent LP problem that can be solved efficiently by the standard simplex method (H .J. Zimmermann 1997).

3.4.3 Problem formulation

This chapter chose the multi-objective functions for solving the TPD problem by reviewing the literature and considering practical situations.(Diaz , S.S Chanas et al, 1984) specified three objective functions to minimize the total transportation costs, total delivery time, and total relative time for a TPD problem with fuzzy multiple objectives. Practical TPD problems typically minimized the total production costs, total transportation costs and total delivery time (Li and K. K. Lai at el. 2000) accordingly, two objective functions were simultaneously considered in developing the proposed MOLP model, as follows.

- Minimize total production and transportation costs

$$M_{in} Z_1 \cong \sum_{i=1}^{m} \sum_{i=1}^{n} \left(P_{ij} + C_{ij} \right) Q_{ij} \tag{1}$$

- Minimize total delivery time

$$\mathrm{M}_{in}\mathrm{Z}_2 \cong \sum_{i=1}^{n}\sum_{j=1}^{m} t_{ij} Q_{ij} \tag{2}$$

Where:

Z_1 total production and transportation costs ($)

Z_2 total delivery time (hours)

Q_{ij} units transported from source *i* to destination *j* (units)

P_{ij} production cost per unit from source *i* to destination *j* ($/unit)

C_{ij} transportation cost per unit from source *i* to destination *j* ($/unit)

t_{ij} transportation time per unit from source *i* to destination *j* ($/unit)

The symbol '$\cong$' is the fuzzified version of '=' and refers to the fuzzification of the aspiration levels. In real world TPD problems, the environmental coefficients and operation parameters are usually uncertain because some information is incomplete or unobtainable over the planning horizon. Accordingly, Equations. (1) & (2) are fuzzy with imprecise aspiration levels, and incorporate the variations in the DM's judgments regarding the solutions of fuzzy multi objective optimization problems. For each of the objective functions of the proposed MOLP model, this work assumes that the DM has such imprecise objective as, "the objective functions should be essentially equal to some value". These conflictin objectives are required to be simultaneously optimized by the DM in the framework of fuzzy aspiration levels.

3.4.4 Constraints

Constraints on total supply available for each source *i*

$$\sum_{j=1}^{n} Q_{ij} = S_{i,} i = 1, 2, \cdots\cdots\cdots\cdots\cdots, m \tag{3}$$

- Constraints on total demand for each destination *j*

$$\sum_{i=1}^{n} Q_{ij} = D_{i,} i = 1, 2, \cdots\cdots\cdots\cdots, n \tag{4}$$

- Non – negativity on total constraints on decision variables

$$Q_{ij} \geq 0, i = 1, 2, \cdots\cdots\cdots m, j = 1, 2, \cdots\cdots\cdots n \tag{5}$$

Where S_i denotes the total supply available of source *i* (units), and D_j denotes the total demand of destination *j* (units). This work addresses a practical application of a FGP model for solving the TPD problem with fuzzy multiple objectives. Therefore, the constraints (3) and (4) in the proposed MOLP model are assumed to be crisp. Notably, the MOLP model described above has a feasible solution only if the total supply available at all sources just equals the total demand required at all destinations

3.5 Transport delivery in Dar es Salaam City

The present public passenger transport system is composed of about 7000 buses, about 4000 taxis and a certain number of rickshaws (SUMATRA, Dar es Salaam City, 2009). All of these vehicles are owned and operated by private entrepreneurs (Ibid). The existing public transport buses, so called daladalas, dominate the public transport delivery in the city (SSATP, 2005). The service provided by daladalas is rather hideous. A comprehensive report by (Kanyama *et al.* 2004), for example, evaluates their service poor and chaotic. The article elaborates that daladalas are unscheduled and thus incur long travel times; they have bad seat arrangements, overload passengers, not follow allocated routes, speed and drive recklessly, etc. The report argues also that most daladalas are "not only second hand but third hand or fourth hand or more" that they emit large amount of pollutants and Carbon Dioxide (CO2). A visit to any number of developing cities can quickly reveal the source of customer dissatisfaction with public transport and non-motorized options. Poor transit services in the developing world push consumers to private vehicle options. Public transport customers typically give the following reasons for switching to private vehicles:

i. Inconvenience in terms of location of stations and frequency of service;
ii. Failure to service key origins and destinations;
iii. Fear of crime at stations and within buses;
iv. Lack of safety in terms of driver ability and the road worthiness of buses;
v. Service is much slower than private vehicles, especially when buses make frequent stops;
vi. Overloading of vehicles makes ride uncomfortable;
vii. Public transport can be relatively expensive for some developing nation households;
viii. . Poor quality or nonexistent infrastructure (e.g., lack of shelters, unclean vehicles, etc.)
ix. Lack of an organized system structure and accompanying maps and information make the systems difficult to use; and
x. Low status of public transit services.

However, all of these problems can be rectified within the modest budgets of developing nation municipalities. Cities such as Bogotá (Colombia), Curitiba (Brazil), and Quito (Ecuador) have dramatically improved transit services with simple solutions. In each case, the city relied upon low cost improvements in public transit and non motorized infrastructure rather than expensive tailpipe technologies. Figures below illustrates the poor quality of public transport in developing cities creates great hardship for the citizen. Figure 3.1 indicate effects of pollution and noise from traffic combustion of fossil fuels produces a number of substances that directly impact upon human health. In the ideal situation, the only result of such combustion is water vapor and carbon dioxide, neither damaging to human health, although carbon dioxide is the main greenhouse gas with impacts on the global climate. However, in reality, combustion is most often not complete and results in the production of substances such as carbon monoxide (CO) and particles (a basic component of particulate matter). Other pollutants due to incomplete combustion processes include volatile organic compounds (VOCs), oxides of nitrogen (NOx) and sulphur dioxide (SO2). If lead has been added to the fuel, lead aerosols are also produced. These by products from combustion, apart from damaging human health, also react in the environment, producing secondary transport pollutants such as sulphuric acid, sulphates and ozone. Atmosphere and climate, together with urban form, population and street density, influence the extent to which populations are exposed to primary and secondary pollutants.

Fig. 3.1. Emission is a threat to peoples' health.

Figure 3.2 indicate overloading in the buses was a problem for people as it often led to incidents of pick - pocketing, impaired air circulation, and bad smells due to warm weather and sweat. The respondents were also worried that overcrowding in the buses could lead to the spread of communicable diseases such as Tuberculosis (TB). Furthermore, overcrowding and squeezing in the buses led to incidents of women being sexually abused by men. When scrambling to enter in the buses become extreme, it is possible to see commuters entering in the buses through the windows. In general, overloading of the buses creates hard travelling conditions for parents with children, women, disabled people and the elderly

Fig. 3.2. Survival of the fittest: Commuters board a bus through the windows during rush hour.

Figure 3.3 indicates that small buses (daladalas) are the most common mode of motorized public transport for households, irrespective of income, in Dar es Salaam. The most dominant types of buses with a capacity of transporting 15 passengers

Fig. 3.3. A series of small buses with capacity of 15 passengers and uncomfortable sitting arrangement.

3.6 Transport balance in the life of the city

There is a high dependence on public transport and also walking is a major mobility function across communities in Dar es Salaam. It follows accordingly that, in order to improve the quality of life of its citizens, the city's development must cater in a large way for these two modes of transport. A major risk or a challenge to the city is the explosion in private car ownership and use generated by incomes increases. A large scale rise in car use demands heavily on public resources to cater for the necessary infrastructure such as roads, but a demand frequently proven to be beyond the capacity. Furthermore, increased pollution, congestion, wide traffic thorough fares and the imposition of cars on walking and living spaces will develop an undesirable living environment. Dar es Salaam has the opportunity to avoid many pitfalls encountered by other developed cities and the initiative already taken to prioritize public transport through the BRT system, which is a major positive step in defining a balanced city. This balance involves applying the right priorities between personal mobility (walking, NMT, cars); the essential movement of goods and freight (port and service vehicles); and an orderly planned public transport system (BRT and associated bus networks). Ultimately, the city is best served if it can create livable communities through a sustainable transport system that enhances and empowers its communities.

3.6.1 Dar es Salaam City towards the world city concept

World cities emerged from the globalization of trade, commerce, and leisure are driven by on line communication and computing technologies and are undoubtedly a "new" breed of cities of which the characteristics surpass these of the well know "mega - city". According to this service based approach, there are three (3) groups of World Cities, complemented by three (3) groups of cities evolving towards becoming a world city (Table 3.1). In the service based classification for world class status, consideration is given to the global capacity of cities in terms of selected services they provide. Global capacity is defined empirically (calculated) in terms of aggregate scores and interpreted theoretically as concentrations of expertise and knowledge. The focus for this classification was on four key services: accounting, advertising, banking and law although other economic activities can also be considered. Cities are evaluated as global service centres in each of these sectors and aggregation of these results to other "supporting" domains provides a measure of a city's global capacity.

Established World Cities				
Classification		**ALPHA CITY (α city)**	**BETA CITY (β city)**	**GAMMA CITY (Γ city)**
Description		*Full service world cities*	*Major world cities*	*Minor world cities*
Examples	*First level*	London; New York; Paris; Tokyo	San Francisco, Sydney, Toronto, Zurich	Amsterdam, Boston, Caracas, Dallas, Düsseldorf, Geneva, Houston Jakarta, **Johannesburg**, Melbourne, Osaka, Prague, Santiago, Taipei, Washington
	Second level	Chicago, Frankfurt, Hong Kong, Los Angeles, Milan, Singapore	Brussels, Madrid, Mexico City, Sao Paulo	Bangkok, Beijing, Montreal, Rome, Stockholm, Warsaw
	Third level		Moscow, Seoul	Atlanta, Barcelona, Berlin, Budapest, Buenos Aires, Copenhagen, Hamburg, Istanbul, Kuala Lumpur, Manila, Miami, Minneapolis, Munich, Shanghai
Emerging World Cities				
Relative strong evidence	Athens, Auckland, Dublin, Helsinki, Luxembourg, Lyon, Mumbai, New Delhi, Philadelphia, Rio de Janeiro, Tel Aviv, Vienna.			
Some evidence	Abu Dhabi, Almaty, Birmingham, Bogota, Bratislava, Brisbane, Bucharest, Cairo, Cleveland, Cologne, Detroit, Dubai, Kiev, Lima, Lisbon, Manchester, Montevideo, Oslo, Riyadh, Rotterdam, Seattle, Stuttgart, The Hague, Vancouver, Ho Chi Minh City.			
Minor evidence	Adelaide, Antwerp, Arhus, Baltimore, Bangalore, Bologna, Brasilia, Calgary, Cape Town, Colombo, Columbus, Dresden, Edinburgh, Genoa, Glasgow, Gothenburg, Guangzhou, Hanoi, Kansas City, Leeds, Lille, Marseille, Richmond, St Petersburg, Tashkent, Tehran, Tijuana, Turin, Utrecht, Wellington.			

Source: J.V. Beaverstock, R.G. Smith and P.J. Taylor. *A Roster of World Cities*, in Cities, 16 (6), (1999), pp 445-458

Table 3.1. Classification of principal World Cities.

The advantage of the "producer service approach" is, according to (Sassen, 1994) that the ranking firmly associates the cities with their tendency to engage with the internationalization, concentration, and intensity of producer services in the world economy. According to the classification methods above, London, New York, Paris, and Tokyo are prime examples of world cities and are also mega cities. However, it is possible and even common for cities that are not mega cities to be world cities and vice versa, overall a world city, also known as a "world class city" can be defined by ten characteristics and the level to which the city incorporates these characteristics determines its status as world city:

i. Name familiarity where the city name is sufficient and there is no need to add the country name.
ii. Active influence and participation in international events and world affairs, with the city housing international headquarters such as the UN (New York), the EU Commission (Brussels) or the European Central Bank (Frankfort).
iii. A fairly large population with at least one million inhabitants but typically several 16 million.
iv. A major international airport acting as high profile hub for several international airlines.
v. An advanced transportation system offering multiple modes of (public) transportation and a highly developed road network.
vi. Home to international cultures and communities or a city which attracts large foreign businesses and related expatriate communities.
vii. Home to international business and stock exchanges that influence the world economy.
viii. Advanced communications infrastructure with WIFI and high--speed broadband.
ix. World renowned cultural institutions and events and a lively cultural scene, including festivals, premieres, music, opera, and theatre scene.
x. Several powerful and influential media outlets with an international reach

3.7 Development frame work, integrated approach and economic growth in Dar es Salaam City

The vision for Dar es Salaam city presents the long term perspective and framework for the Dar es Salaam Transport Policy and System Development Master Plan. It discusses both the general development vision for Dar es Salaam with the year 2030 perspective and the more specific and directly inters - related vision for the future transport system in Dar es Salaam. The year 2003 National Transport Policy underwrites the principle of achieving "... efficient and cost effective domestic and international transport services to all segments of the population and sectors of the national economy with maximum safety and minimum environmental degradation." Realizing the principle requires substantial efforts oriented to creating "...safe, reliable, effective, efficient and fully integrated transport infrastructure and operations which will best meet the needs of travel and transport at improving levels of service at lower costs in a manner, which supports government strategies for, socio - economic development whilst being economically and environmentally sustainable." The transport policy thus proclaims that initiatives taken in its policy, economy, society and environment will define the city's long - term development. Figure 3.4 below is the development frame work of Dar es Salaam city. The "National Transport Policy and System Development Master Plan" argues that the fundamental requirement for improvement is the establishment of an adequate institutional framework. Creating the appropriate institutional framework cannot be realized successfully without a balanced and integrated approach where vision, strategy, and action are intertwined and part of a wider vision for sustainable economic development (Figure 3.5).

Fig. 3.4. Developing frame work.

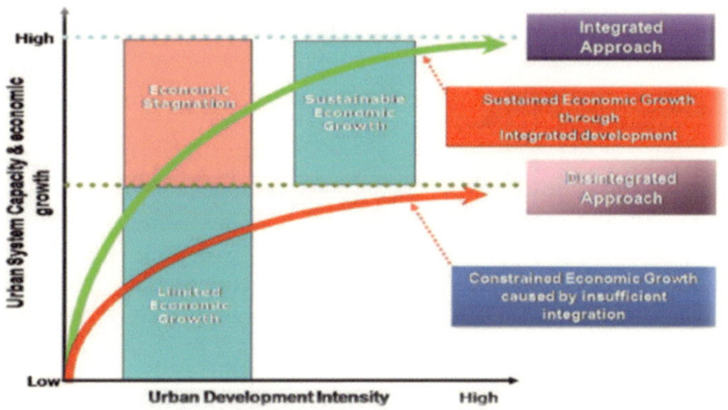

Fig. 3.5. Integrated approach and economic growth.

The National Transport Policy acknowledges the need for an integrated approach when it links the efforts in the field of transport with the long term development goal of the country as expressed in other national guidelines such as the National Poverty Reduction Strategy Paper (PRSP), the Rural Development Strategy (RDS), the Civil Service Reform Program, and reform programs aiming at private sector involvement in economic development, strategic environmental sustainability, gender issues, eradication of diseases and literacy campaigns. The National Transport Policy therewith recognizes the pivoting role of transport and acknowledges that the realization of objectives in "...priority sectors such as education, health, water, agriculture, manufacturing, tourism, mining, energy, land and good governance hinges on the availability of adequate and reliable transport to reach inputs to production points and also to distribute outputs from production points to consumption points/markets. The integrated approach is thus imperative and finds justification in the scale and scope of the urban transport problem including: "... high cost, low quality services due to various reasons including the existence of high backlog of infrastructure maintenance and rehabilitation, inadequate institutional arrangements, laws, regulations and procedures which are not consistent or compatible with each other to create conducive climate for investment and hence growth of the sector, inadequate capacity caused by low level of investment in resources, and low level of enforcement of safety, environmental sustainability and gender issues. The "Transport Policy and System

Development Master Plan" therefore entrenches the future transport system for Dar es Salaam into a wider vision related to the long term social and economic development of the City. The Dar es Salaam Development Vision follows a sequential logic where the *Vision* leads to a *Strategy*, itself defined by a number of concrete *Actions* of which the development of an integrated urban transport system is one, albeit critical, component (Figure 3.6). The approach for the transport policy and system development master plan follows a structured and hierarchical (sequential) approach:

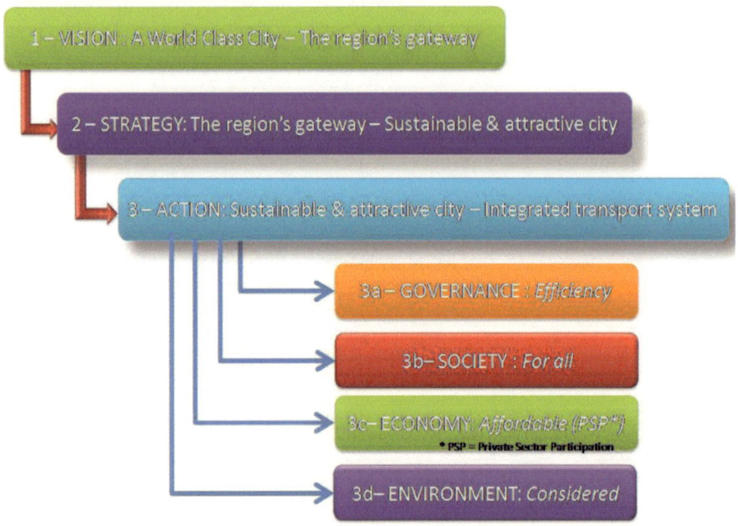

Fig. 3.6. Integrated approaches for the new urban transportation system.

i. The vision is a long term and final objective according to which all initiatives need to be targeted. Achieving this vision might take several decades but it provides a general framework along which strategies can be / are developed that contribute to proceeding towards the realization of the vision
ii. The strategy for Dar es Salaam is the translation of the vision into concrete initiatives in different areas that contribute to the realization of the vision (the growth and development of Dar es Salaam towards becoming a world class city). The key of the strategy is that all actions (sub strategies) need to work towards the transformation of the city into a sustainable and attractive city.
iii. The Actions necessary to create a sustainable and attractive city: Actions are needed in different areas, such as urban planning, land use planning, environmental protection, etc. One of the critical areas for concrete action is the (long - term) urban transport strategy for which the "Transport Policy and System Development Master Plan" will formulate the "Dar es Salaam Transport Vision - 2030" which provides the four (4) building blocks essential to achieving attractiveness and sustainability of the city, namely:
    a. *Governance*: considering the most efficient structure to develop and manage an integrated transport system, now and in the future;
    b. *Society*: ensuring that the transport system is accessible to all, including (and in particular) to the less fortunate and the poor members of the community as well as

the physically challenged persons. At the same time, it should be of a quality that is attractive and acceptable for visitors, in particular tourists and business persons.

c. *Economy*: where realism on the affordability of the transport system should be the guiding principle which means that infrastructure planning and transport services should consider budgetary constraints, not only focusing capital investments but equally and maybe even more importantly considering long term maintenance and operating costs. To increase the possibilities of capital expenditure for major infrastructure developments, a gradual reduction of the present dependence up on international donors and IFI's should be pursued by inviting the private sector to participate in the efforts.
d. *Environment*: where the (long term) impact on the environment, in particular on fauna and flora and on the quality of air should be considered important elements of sustainability and attractiveness of the city. Lack of fauna and flora is unattractive and creates "unsafe" population concentrations while a poor quality of air impacts the health of its inhabitants (e.g., respiratory problems). All these problems will increase the cost of social well being and will put pressure on the city's public budget and spending capacity.

The "Transport Policy and System Development Master Plan for Dar es Salaam" thus should recommend concrete actions for the creation of an integrated transport system in the city. The action plan will be embedded in the comprehensive strategy to create a sustainable and attractive city, based upon a clear vision about that future development of Dar es Salaam. The figures 3.7 and 3.8 below illustrate the dual dimension of the city plan and integrated transportation system respectively.

Fig. 3.7. The dual dimension of the Transport Plan.

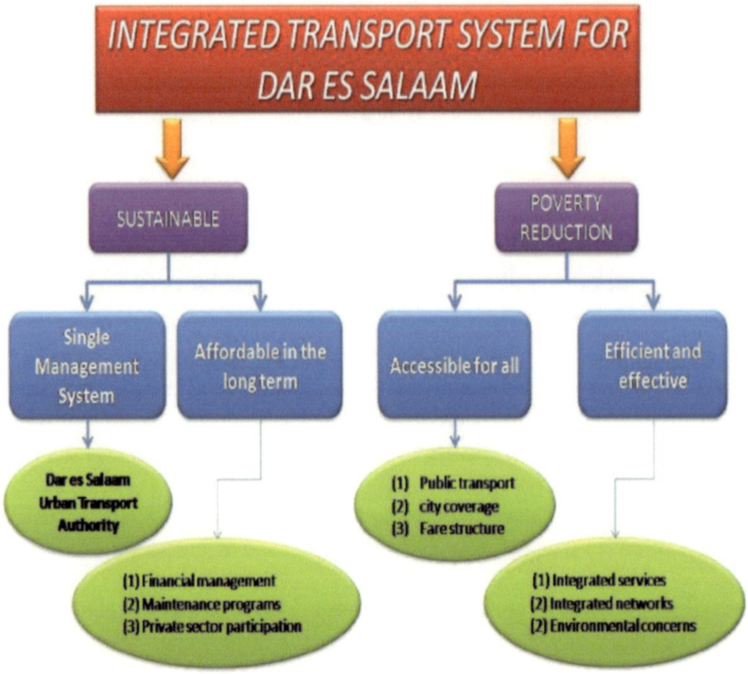

Fig. 3.8. Integrated transportation system.

3.8 General urban transport enhancement strategy

Transport is an entitlement to the citizens of Dar es Salaam, and good transport networks have multiplicity of benefits; *socially, economically, environmentally and culturally*. Equity is also important as transport should be affordable to all so that there is equal and affordable access to opportunities of employment, education and social inclusion. Furthermore the paybacks in both direct and indirect terms are substantial and will contribute directly to improved economic performance, productivity and greatly reduce the negative stresses that citizens endure on a daily basis. The increasingly difficult urban transport situation in the city, characterized by a high degree of traffic congestion, constrained resources for urban transport and deteriorating air quality, lies in the forefront of concerns. Urban transport problems are borne out of a set of complex and diverse environmental and economic factors and profound institutional failures. In Dar es Salaam, due to a low level of car ownership and high dependence on public transport, the problems of public transport are synonymous with the problems of urban transport because public transport vehicles (Dala Dala) serve such a large proportion of total trip demand. The present public transport system in Dar es Salaam is highly unsatisfactory from the perspectives of all stakeholders: the public, the city, the operators and the users. Government is now addressing its growing transport problems with the introduction of the Bus Rapid Transit (BRT) with the essential associated administrative and institutional reforms. A key consideration in this regard is that, most certainly within the near term planning horizon, the need to move people must take precedence over the need to move vehicles. However, in the medium to longer term, other transport pressure will arise. If history holds any lessons, it is that future growth in income will inevitably catalyze an increase in trip making, as well as changes in the modes used to accomplish such trips. It is likely that private modes of transport, such as passenger cars, will continue to become increasingly popular with the citizens of Dar es Salaam. It is expected that the current 74,000 private vehicles located within the metropolitan area will increase to about 180,000 by year 2015, and near 515,000 by year 2030 (JICA 2008). This means that vehicle ownership will more than triple from 25 cars per 1,000 persons in year 2007, to 89 cars per 1,000 persons by year 2030. Pronounced impacts on Dar es Salaam congestion, and the need for additional road infrastructure, are consequently expected. The construction of BRT will certainly be a key mitigating factor in defining the modal choice relationship in terms of inducing mode switching and providing increased mobility for transit dependent elements of the population. However, demand on the road network is nevertheless expected to dramatically increase in future in line with rising socio economic well being of the populace. No single remedy can be expected to comprehensively address such phenomena, instead, a more holistic approach is needed which relies on intermodality and a harmonious combination of the various modes that compromise a multi faceted and integrated urban transport system. Herein lies the challenge; solutions are needed whose practicality can be viewed through the prism of existing realities, and whose validity will remain intact over the Master Plan planning horizon extending to year 2030. In defining transport systems for the future, the chapter has fully considered a number of key features which will dictate the nature of transport evolution from a strategic viewpoint; namely

i. The physical characteristics (space for road expansion, increasing traffic etc) and urban structure (type and extent of land use) of the city;
ii. A developing city with constrained financial resources;
iii. The social characteristics of its people (high dependency on public transport, need for mobility to increase opportunity and reduce poverty; increasing well being in future); and
iv. The policy and regulatory frameworks as key to developing sustainable transport.

One of the key recommendations in this chapter is that, in order to improve what is at present an overlapping and (often) ineffective organizational approach to developing road systems, executing traffic control and management, as well as operating public transport, the formation of a multi disciplinary and multi modal Dar es Salaam Urban Transport Authority (DUTA). At the same time capacity development is very necessary in the field of transport planning and administration. Also an organization named National Center for Transport Studies (NCTS), whose responsibility is placed on education, research and development is proposed.

4. Intelligent transportation system innovations

Overview of the widespread application of ITS has been anticipated for some years and since the early 1990s, there have been numerous model based attempts to predict their impact. ITS is not just about the provision of information to travelers. It can affect the nature of the travel experience (e.g., by simplifying the process of paying fare or road toll, by enabling the driver to pre book his parking space or conduct business by phone while en route to the office, or by taking control of the vehicle in hazardous situations). It can also extend the range of levers available to the system manager (e.g., by making it possible to charge motorist that reflect the current level of congestion or pollution, or to detect and prosecute a wide range of traffic violations). All of these cold influence traveler behavior in various ways, for example, recognize that mode choices might be affected by the provision of simplified ticketing or tolling system or by the ability to work end route, and that the choice of car park might be strongly influenced by the possibility of pre booking space in some location and so on.

4.1 Intelligent transportation system technologies applied to BRT system in developing countries

ITS technologies are being implemented more commonly in European, North American countries, and Australia than in developing countries. BRT systems in developing countries are still limited in ITS applications because of the capital and operating costs (Wright 2004). ITS technologies mainly contribute to the image, safety, and operating speed (Kittelson & Associates et al. 2003, Darido et al. 2006, Currie 2006 and Sakamoto et al. 2007) but are not essential features for a successful BRT system. The BRT systems in Bogotá, Quito, Beijing, Mexico City, and all Brazilian systems are successful examples that have not implemented or have very limited ITS technologies. Transit Signal Priority (TSP), real time passenger information systems, and Automatic Fare Collection (AFC) are examples of typical ITS applications in BRT systems. Implementation of TSP has grown rapidly among the U.S. transit systems. Real time passenger information systems increase productivity of passengers while waiting for buses, avoid crowding at stations, and enhance the image of the shelters (Kittelson & Associates et al 2003). Automatic Vehicle Location (AVL) systems help track the locations of vehicles, which can be used for real time fleet management and future planning purposes. The global positioning system based AVL system is perhaps the most popular among the available location technologies (Gillen and Johnson 2002). One of the new ITS technologies for BRT are lane assist systems being implemented in the BRT systems in Orlando and Minneapolis. Lane assist permits BRT vehicles to operate at higher operating speeds with improved safety (Kulyk and Hardy 2007). Precision docking technology (implemented in Las Vegas, but more popular in European cities) helps reduce dwell time. Some features below applied in several BRT systems.

i. *Fare Collection Methods* - Automatic fare collection (AFC), although originating in other transit systems, has become a regular feature of BRT systems worldwide. Advanced AFC with a common smart card allows integration of several modes in one single system, which offers customer convenience (GTZ 2006). In surveys carried out among transit users in Hong Kong, Taipei, New Delhi, London, Oslo, Copenhagen, Washington D.C., San Francisco, Chicago, Rome, Bangkok, Seoul, and Istanbul, smart cards were noted as being effective in promoting ridership, increasing customer satisfaction, improving boarding time, and increasing ease of access (Boushka 2006). AFC usually generates important data for demand forecasting and operational planning (Hidalgo et al. 2007). However, three recent examples demonstrate that AFC may not be as beneficial as it appears. The first example is AFC on the Silver Line in Boston. AFC equipment initially was implemented with the purpose of saving running time. However, contrary to expectations, the travel time increased after AFC implementation. Such experience illustrates the importance of dwelling time control (Darido et al. 2006). The second and third examples are the Quito and Jakarta BRT systems, where the implementation time for user adaptation to AFC technology has been considerably short, causing "insufficient testing and quality assurance." In addition, their fare collection systems are not compatible with other public transportation modes or even among different BRT corridors in the same city (Hidalgo et al. 2007).
ii. *Operating Speed* - Operating speed depends on many factors such as guide ways, number of stops, dwell time, etc. When Bogotá's TransMilenio was first implemented,

the operating speed went from approximately 15 km/h to 26.7 km/h (Cain 2007). In Seoul, the operating speed of buses has improved after the implementation of BRT in 2004 (by 2.7 km/h to 11 km/h, depending on the corridor), and the speed has increased as users become more familiar with the system (GTZ 2006). Operating speed has a direct impact on ridership attraction. As the name implies, BRT service should be "rapid."The travel time and ridership attraction of the BRT features reviewed above, all are aimed at reducing travel time or increasing ridership. Therefore, travel time savings (for users) and ridership attraction (for agencies and operators) are the most important design goals. In fact, the most distinctive features of BRT systems are the ones that contribute most to reduction in travel time (such as guide ways, high capacity vehicles, high service frequency, TSP, AFC) and ridership attraction (such as enhanced stations and shelters, transit oriented development, real time passenger information systems, route coverage).

4.2 Bus rapid transit system creating better mobility

BRT is an enhanced bus system that operates on bus lanes or other transit ways in order to combine the flexibility of buses with the efficiency of rail. By doing so, BRT operates at faster speeds, provides greater service reliability and increased customer convenience. It also utilizes a combination of advanced technologies, infrastructure and operational investments that provide significantly better service than traditional bus service. Bus systems provide a versatile form of public transportation with the flexibility to serve a variety of access needs and an unlimited range of locations throughout a metropolitan area. Because buses travel on urban roadways, infrastructure investments needed to support bus service can be substantially lower than the capital costs required for rail systems. As a result, bus service can be implemented cost effectively on routes where ridership may not be sufficient or where the capital investment may not be available to implement rail systems. Traffic congestion, urban sprawl, central city decline, and air pollution are all problems associated with excessive dependence on automobiles. Increasing recognition of the need for high quality transit service to alleviate these conditions has fueled growing demand for new public transportation service. Despite the inherent advantages of bus service in terms of flexibility and low capital cost, the traveling public frequently finds the quality of bus service provided in urban centers to be wanting. Conventional urban bus operations often are characterized by sluggish vehicles inching their way through congested streets, delayed not only by other vehicles and traffic signals, but also by frequent and time consuming stops to pick up and discharge passengers. Buses travel on average at only around 60 percent of the speeds of automobiles and other private vehicles using the same streets due to the cumulative effects of traffic congestion, traffic signals, and passenger boarding. Moreover, compared to rail systems, the advantageous flexibility and decentralization of bus operations also result in a lack of system visibility and permanence that contributes to public perceptions of unreliability and disorganization.

4.2.1 What is bus rapid transit?

Low cost investments in infrastructure, equipment, operational improvements, and technology can provide the foundation for *Bus Rapid Transit* systems that substantially upgrade bus system performance. Conceived as an integrated, well defined system, Bus Rapid Transit would provide for significantly faster operating speeds, greater service reliability, and increased convenience, matching the quality of rail transit when implemented in appropriate settings. Improved bus service would give priority treatment to buses on urban roadways.

4.2.2 Why bus rapid transit?

Transportation and community planning officials all over the world are examining improved public transportation solutions to mobility issues. This renewed interest in transit reflects concerns ranging from environmental consciousness to the desire for alternatives to clogged highways and urban sprawl. These concerns have led to a re - examination of existing transit technologies and the embrace of new, creative ways of providing transit service and performance. BRT can be an extremely cost effective way of providing high quality, high performance transit. Advancements in technology such as clean air vehicles, low floor vehicles, and electronic and mechanical guidance

4.2.3 Overview of bus rapid transit system

With population increase, increasing transportation demand has lead to debilitating traffic in most major cities. Figure 4.1 indicate population increases of some major cities.

Bogota - Colombia California - USA

Fig. 4.1. Population increase of some major cities.

In order to address this situation, a high pax capacity transport alternative must be chosen system pax/hour/direction capacity ('000 pax

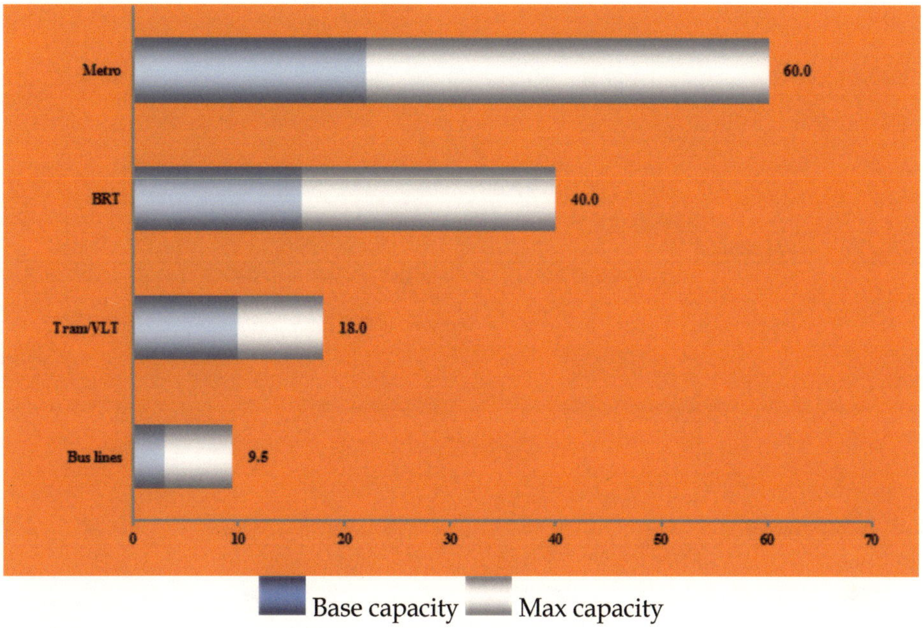

Fig. 4.2. Different mode of public transportation capacity.

In comparison to rail based systems, a BRT system provides a high pax system with a significantly lower implementation cost and time. Figure 4.3 and 4.4 indicates the different mode of transportation implementation cost (USSMM/KM) and time (Mths) respectively.

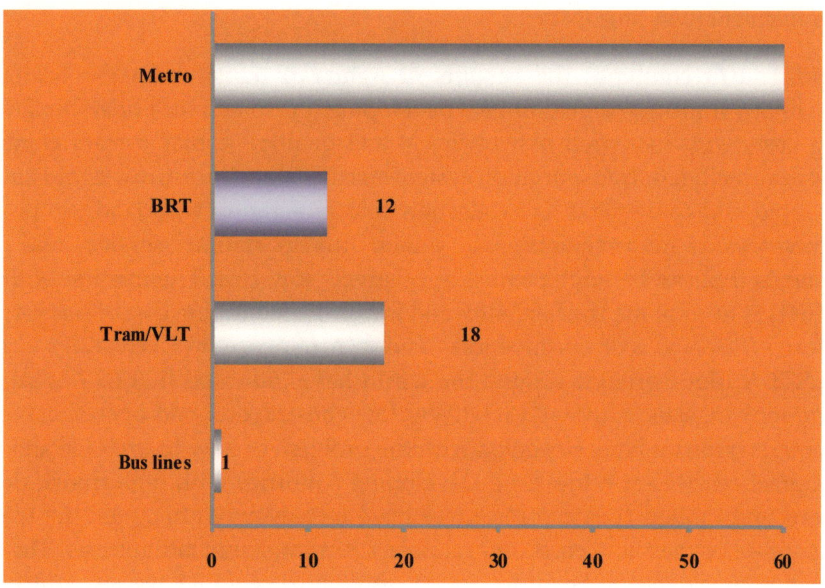

Fig. 4.3. Implementation cost (US$ MM/.

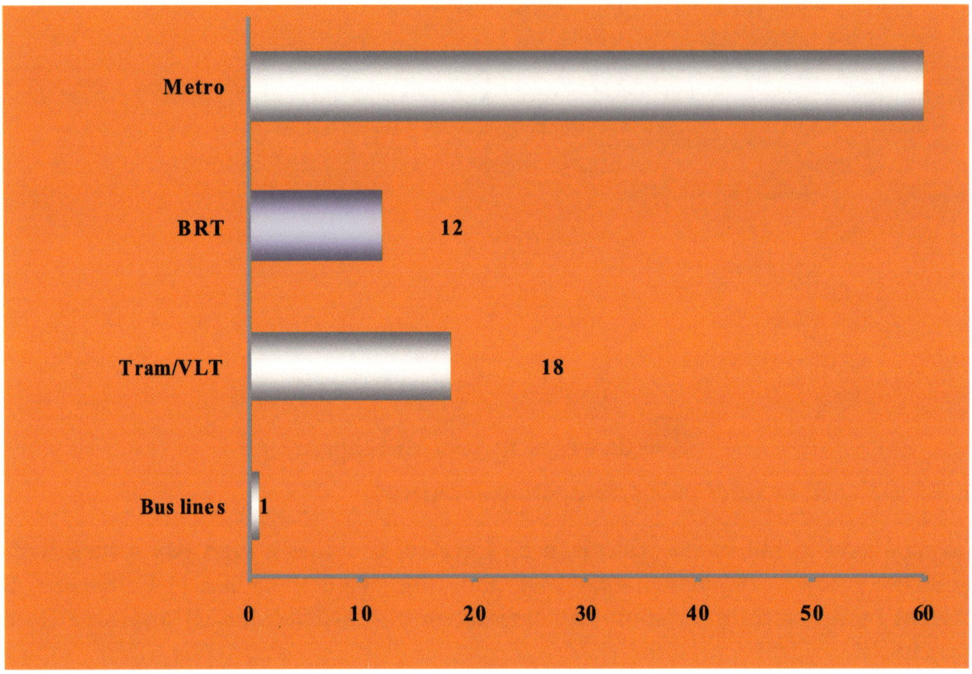

Fig. 4.4. Implementation time (Mths).

Bus Rapid Transit combines the benefits of light rail transit with the flexibility and efficiency of bus transit. The goal of BRT development is to enhance ridership and reduce operating costs with increased service levels and quality. A BRT system combines the technology of intelligent transportation systems, traffic signal priority, cleaner and quieter vehicles, rapid and convenient fare collection, and integration with land use policy. BRT has demonstrated improvements in public transportation service and enabled improvements that can be implemented at relatively low cost. The mission of Bus Rapid Transit (BRT) is to combine the flexibility and low implementation cost of bus service with the comfort, efficiency, cost effectiveness, land use influence and versatility of light rail transit (LRT). Various projects around the world have indicated that BRT is an effective alternative for congested cities at a relatively low construction and operation cost. Cities in developing countries have struggled with the problem of how to upgrade and improve existing transit services at a low cost. Developing countries with high transit dependent populations and limited financial resources have increasingly attempted the use of BRT systems because of their low costs and relatively fast implementation times. The cost of a BRT project is considered to be approximately one third of a LRT project, which is a cost that developing countries can afford. After construction the system is practically self financing with fares of about $US0.50 per trip. BRT has proved that it allows low fares and reduced travel times for low income users. BRT systems such as Curitiba in Brazil and Transmilenio in Colombia are great examples of the success that the BRT system has had in Latin American countries.

4.2.4 BRT – Concepts and evolution

There is a broad range of perspectives as to what constitutes BRT. At one end of the spectrum, BRT has been defined as a corridor in which buses operate on a dedicated right of way such as a bus way or a bus lane reserved for buses on a major arterial road or freeway. Although this definition describes many existing BRT systems, it does not capture the other features that have made rail rapid transit modes so attractive around the world. BRT has also been defined as a bus based rapid transit service with a completely dedicated right of way and on line stops or stations, much like LRT. This is consistent with the FTA definition of BRT as "a rapid mode of transportation that can combine the quality of rail transit and the flexibility of buses". For the purpose of this chapter, BRT has been defined more comprehensively as a flexible, rubber tired form of rapid transit that combines stations, vehicles, services, running ways, and ITS elements into a fully integrated system with a strong image and identity. BRT applications are designed to be appropriate to the market they serve and their physical surroundings, and they can be incrementally implemented in a variety of environments (from rights of way totally dedicated to transit to streets and highways where transit is mixed with traffic). In brief, BRT is a fully integrated system of facilities, services, and amenities that are designed to improve the speed, reliability, and identity of bus transit. In many respects, it is rubber tired LRT, but with greater operating flexibility and potentially lower capital and operating costs. Often, a relatively small investment in dedicated guide ways can provide regional rapid transit. This definition has the following implications:

- Where BRT vehicles (buses) operate totally on exclusive or protected rights of way (surface, elevated, and/or tunnel) with on-line stops, the level of service provided is similar to that of heavy rail rapid transit (metros).
- Where buses operate in combinations of exclusive rights of way, median reservations, bus lanes, and street running with on line stops, the level of service provided is similar to that of LRT.
- Where BRT operates almost entirely on exclusive bus or HOV lanes on highways (freeways and expressways) to and from transit centers with significant parking and where it offers frequent peak service focused on a traditional CBD, it provides a level of service very similar to that of commuter rail.
- Where buses operate mainly on city streets with little or no special signal priority or dedicated lanes, the level of service provided is similar to that of an upgraded limited stop bus or tram system.

Figure 4.5 describes the seven major components of BRT running ways, stations, vehicles, service, route structure, fare collection, and ITS. Collectively, these components form a complete rapid transit system that can improve customer convenience and system performance.

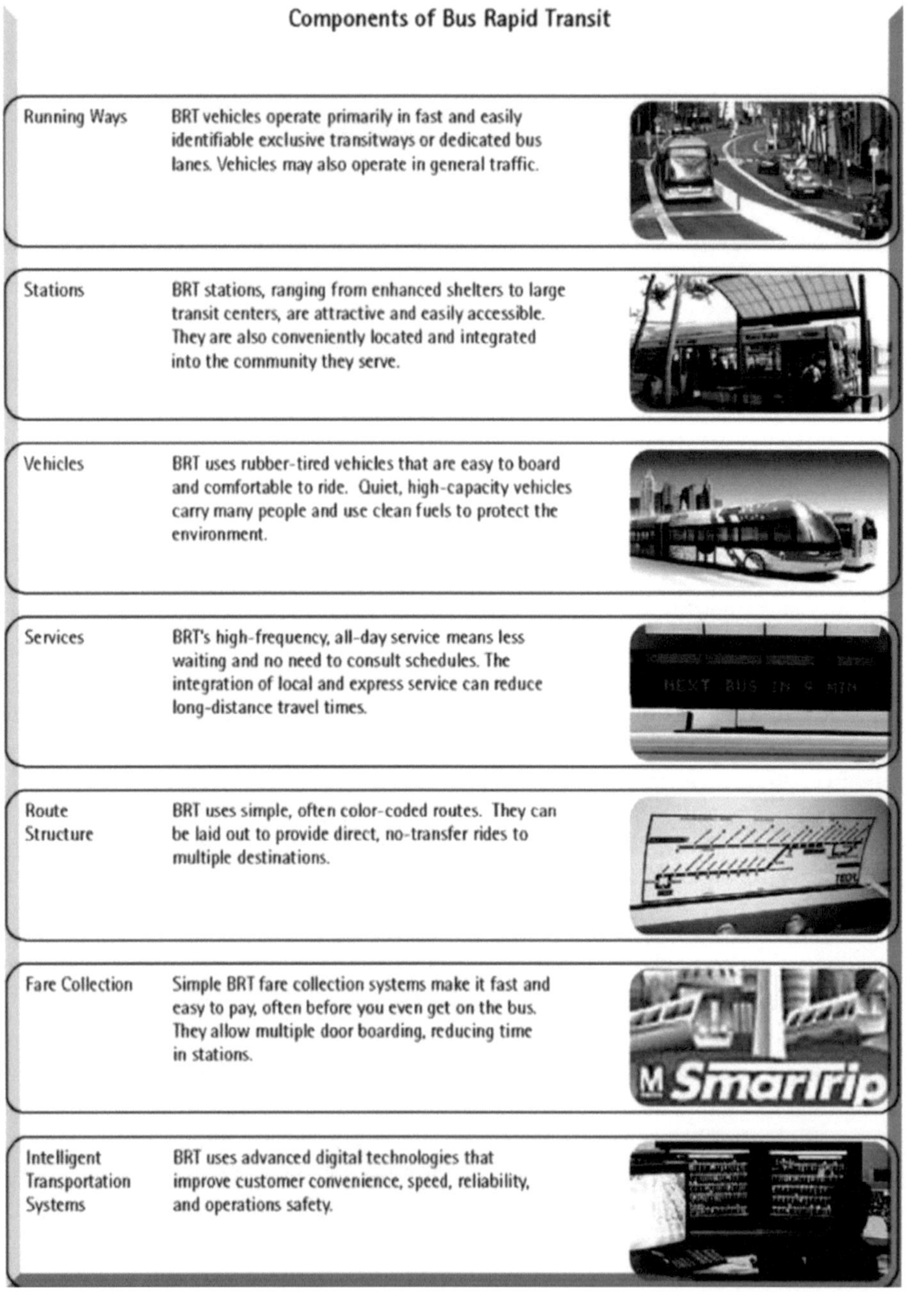

Fig. 4.5. Major components of BRT.

4.2.5 Limitation of BRT

However, effectiveness of BRT is not always permanent. (Vuchic, 2005) pointed out that BRTs cannot succeed if police enforcement is not strict due to security system, citing the examples of Philadelphia and Mexico. Experiences from the U.S. cities such as Shirley Bus way in Washington and El Monte Bus way in Los Angeles show that pressures by automobile interests are threat to the existence of BRT. Relationship of BRT to other modes is a crucial factor for the success of BRT: BRT cannot bring success as a standalone policy and effectiveness depends on the presence of complementary transport options, such as promotion of non motorized transport and integrated feeder services (Wright, 2001). Another important factor for success understanding of planning and design elements, based on experiences in real world conditions (Vuchic, 2005).

4.3 Experience of BRT System as a mode of public transport from other cities

4.3.1 The BRT system in Curitiba – Brazil

The bus system of Curitiba, Brazil, exemplifies a model Bus Rapid Transit system, and plays a large part in making this a livable city. The buses run frequently some as often as every 90 seconds and reliably, commuters ride them in great numbers, and the stations are convenient, well designed, comfortable, and attractive. Curitiba has one of the most heavily used, yet low cost, transit systems in the world. It offers many of the features of a subway system vehicle movements unimpeded by traffic signals and congestion, fare collection prior to boarding, quick passenger loading and unloading but it is above ground and visible. Even with one automobile for every three people, one of the highest automobile ownership rates in Brazil, and with a significantly higher per capita income than the national average, around 70 percent of Curitiba's commuters use transit daily to travel to work. Greater Curitiba with its 2.2 million inhabitants enjoys congestion free streets and pollution free air.

4.3.1.1 Evolution of the bus system in Curitiba

The bus system did not develop overnight, nor was it the result of transit development isolated from other aspects of city planning. It exists because thirty years ago Curitiba's forward thinking and cost conscious planners developed a Master Plan integrating public transportation with all elements of the urban system. They initiated a transportation system that focused on meeting the transportation needs of the population rather than focusing on those using private automobiles and then consistently followed through over the years with staged implementation of their plan. They avoided large scale and expensive projects in favor of hundreds of modest initiatives. A previous comprehensive plan for Curitiba, developed in 1943, had envisioned exponential growth of automobile traffic and wide boulevards radiating from the central core of the city to accommodate the traffic. Rights of way for the boulevards were acquired, but many other parts of the plan never materialized. With the adoption of the new Master Plan in 1965, the projected layout of the city changed dramatically. The Master Plan sprang from a competition among urban planners prompted by fears of city officials that Curitiba's rapid growth, if unchannelled, would lead to the congested, pedestrian unfriendly streets and unchecked development that characterized their neighbor city, São Paulo, and many other Brazilian cities to the north. As a result of the Master Plan, Curitiba would no longer grow in all directions from the core, but would grow along designated corridors in a linear form, spurred by zoning and land use policies promoting high density industrial and residential development along the corridors. Downtown Curitiba would no longer be the primary destination of travel, but a hub and terminus. Mass transit would replace the car as the primary means of transport within the city, and the high density development along the corridors would produce a high volume of transit ridership. The wide boulevards established in the earlier plan would provide the cross section required for exclusive bus lanes in which express bus service would operate.

4.3.1.2 The Curitiba's bus system

Curitiba's bus system evolved in stages over the years as phases of the Master Plan were implemented to arrive at its current form. It is composed of a hierarchical system of services. Small minibuses routed through residential neighborhoods feed passengers to conventional buses on circumferential routes around the central city and on interdistrict routes. The backbone of the bus system is composed of the express buses operating on five main arteries leading into the center of the city much as spokes on a wheel lead to its hub. This backbone service, aptly described as *Bus Rapid Transit*, is characterized by several features that enable Curitiba's bus service to approach the speed, efficiency, and reliability of a subway system: integrated planning ; exclusive bus lanes ; signal priority for buses ; Pre - boarding fare collection ; level bus boarding from raised platforms in tube stations ; free transfers between lines (single entry) ; large capacity articulated and bi-articulated wide door buses ; and overlapping system of bus services.

Each artery is composed of a "trinary" road system, consisting of three parallel routes, a block apart. The middle route is a wide avenue with "Express" bus service running down dedicated high capacity express bus ways in the center two lanes, offering frequent stop service using standard, articulated and bi - articulated buses carrying up to 270 passengers a piece. The outer lanes are for local access and parking. Back in the 1960s the building of a light rail system in these avenues had been considered, but proved to be too expensive. The two outer routes are one way streets with mixed vehicle traffic lanes next to exclusive bus lanes running "direct" high speed bus service with limited stops. Both the express and direct services use signal priority at intersections.

Buses running in the dedicated and exclusive lanes stop at tube stations. These are modern design cylindrical shaped, clear walled stations with turnstiles, steps, and wheelchair lifts. Passengers pay their bus fares as they enter the stations, and wait for buses on raised station platforms. Instead of steps, buses are designed with extra wide doors and ramps which extend when the doors open to fill the gap between the bus and the station platform. The tube stations serve the dual purpose of providing passengers with shelter from the elements, and facilitating the efficient simultaneous loading and unloading of passengers, including wheelchairs. A typical dwell time of only 15 to 19 seconds is the result of fare payment prior to boarding the bus and same level boarding from the platform to the bus. Passengers pay a single fare equivalent to about 40 cents (U.S.) for travel throughout the system, with unlimited transfers between buses. Transfers are accomplished at terminals where the different services intersect. Transfers occur within the prepaid portions of the terminals so transfer tickets are not needed. In these areas are located public telephones, post offices, newspaper stands, and small retail facilities to serve customers changing buses. Ten private bus companies provide all public transportation services in Curitiba, with guidance and parameters established by the city administration. The bus companies are paid by the distances they travel rather than by the passengers they carry, allowing a balanced distribution of bus routes and eliminating the former destructive competition that clogged the main roads and left other parts of the city unsaved. All ten bus companies earn an operating profit. The city pays the companies for the buses, about 1 percent of the bus value per month. After ten years, the city takes control of the buses and uses them for transportation to parks or as mobile schools. The average bus is only three years old, largely because of the recent infusion of newly designed buses, including the articulated buses, into the system.

4.3.1.3 Integration of transit with land use planning

Curitiba's Master Plan integrated transportation with land use planning, with the latter as the driving force, and called for a cultural, social and economic transformation of the city. It limited central area growth, while encouraging commercial growth along the transport arteries radiating out from the city center. The city's central area was partly closed to vehicular traffic, and pedestrian streets were created. The linear development along the arteries reduced the traditional importance of the downtown area as the primary focus of day to day transport activity, thereby minimizing congestion and the typical morning flow of traffic into the central city and the afternoon outflow. As a result, during any rush hour in Curitiba, there are heavy commuter movements in both directions along the public transportation arteries. The Master Plan also provided economic support for urban development along the arteries through the establishment of industrial and commercial zones and mixed use zoning, and encouraged local community self sufficiency by providing each city district with its own adequate education, health care, recreation, and park areas. By 1992, almost 40 percent of Curitiba's population resided within three blocks of the major transit arteries. Other policies have contributed to the success of the transit system, in the areas of zoning, housing development, parking and employer paid transit subsidies. Land within two blocks of the transit arteries has been zoned for mixed commercial and residential uses. Higher densities are permitted for office space, since it traditionally generates more transit ridership per square foot than residential space. Beyond these two blocks, zoned residential densities taper with distance from transit ways. Land near transit arteries is encouraged to be developed with community assisted housing. The Institute of Urban Research and Planning of Curitiba (IPPUC), established in the 1960s to oversee implementation of the Master Plan, must approve locations of new shopping centers. They discourage American style auto oriented shopping centers by channeling new retail growth to transit corridors. Very limited and time restricted public parking is available in the downtown area, and private parking is very expensive. Finally, most employers offer transportation subsidies to workers, especially low skilled and low paid employees, making them the primary purchasers of tokens.

4.3.1.4 Staged development of the bus system

As the population increased during the period from 1970 through the present, Curitiba's bus system evolved incrementally. It required expansion of service routes, frequencies, and capacities, and improvements in fare payment, scheduling, and facility design to facilitate the passenger transferring process. Innovative low cost and low tech options for new services and features were chosen over more expensive alternatives at each stage. Planners did not hesitate to abandon choices that did not work in favor of more effective solutions. At several points throughout the bus system development, the option of constructing a rail network was considered. Initially, buses were chosen over rail because they were far more adaptable and cheaper for a developing city such as Curitiba. In the mid 1980s the ridership had grown enough to support a rail network, but capital costs were prohibitive. Instead, the high capacity, high speed service known as "direct" service was eventually introduced on the one way exclusive bus lanes that parallel the main corridors one block away. This service, including the tube stations, cost about $200,000 per kilometer to build, and was far cheaper, faster and less disruptive than the estimated $20 million per kilometer for a light rail system. Not to be underestimated in the evolution of the transit system is the influence of the current governor of the State of Parana, Jaime Lerner. Lerner left his position as president of the IPPUC to become a three--time Mayor of Curitiba, and then governor. With a stake in the development of the Master Plan, he was its champion throughout the years, providing guidance, a firm governmental commitment to transit, and leadership. His steady promotion of the plan enabled it to withstand any tendencies for local politics to alter its course.

4.3.1.5 Results of bus rapid transit

The popularity of Curitiba's Bus Rapid Transit system has affected a modal shift from automobile travel to bus travel, in spite of Curitibanos' high income and high rate of car ownership relative to the rest of Brazil. Based on 1991 traveler survey results, it was estimated that service improvements resulting from the introduction of Bus Rapid Transit had attracted enough automobile users to public transportation to cause a reduction of about 27 million auto trips per year, saving about 27 million liters of fuel annually. In particular, 28 percent of direct bus service users previously traveled by car. Compared to eight other Brazilian cities its size, Curitiba uses about 30 percent less fuel per capita, because of its heavy transit usage. The low rate of ambient air pollution in Curitiba, one of the lowest in Brazil, is attributed to the public transportation system's accounting for around percent of private trips in the city. Residential patterns changed to afford bus access on the major arteries to a larger proportion of the population. Between 1970 and 1978, when the three main arteries were built, the population of Curitiba as a whole grew by 73 percent, while the population along the arteries grew by 120 percent. Today about 1,100 buses make 12,500 trips per day, serving more than 1.3 million passengers per day, 50 times more than 20 years ago. Eighty percent of the travelers use either the express or direct bus service, while only 20 percent use the conventional feeder services. Plans for extending the rapid bus network will reduce the need for conventional services. In addition to enjoying speedy and reliable service, Curitibanos spend only about 10 percent of their income on travel, which is low relative to the rest of Brazil.

4.3.2 The case of transmilenio the BRT system in Bogota – Colombia

4.3.2.1 Before the BRT implementation

This section describes the transportation conditions in Bogotá in 1998, (before Transmilenio) in order to understand the positive transportation changes made in the city. Ninety five percent of the road network was used by 850,000 private vehicles, which transported about 19% of the population. Close to 70% of trips shorter than 3 km were made by car. Buses occupied a low percentage of the roadway network. Seventy two percent of trips were made by public transit on about 21,000 buses. The average trip by bus was about 1 hour 10 minutes in duration with an average speed of 10 km/hr. The majority of the buses were more than 14 years old with an average of 50% occupancy. A total of 48% of public transit vehicles were medium sized buses (40 - 80 passengers), 37% small buses (20 - 40 passengers), and 15% were minibuses. The fares ranged between US$0.30 and US$0.40 depending of the type and age of the buses (Transmilenio S.A., 2000). In general buses did not have comfortable seats, ventilation or security. There were no defined bus stops, therefore, buses picked up and dropped off passenger at any location along the route. There was no motivation for car owners to switch to public transportation because of the low quality of the system. Figure 4.6 shows some of the buses previously used in the public transportation system in Bogotá.

Fig. 4.6. Bogotá, Before the BRT Implementation.

The transportation system was operated by multiple private operators, which perceived their income as a function of the number of buses in their fleet. The bus system growth was very fast and disorganized. Between 1993 and 1997 the demand for bus service increased 27%. On the other hand, the bus supply increased 72%, which shows the lack of control and planning of the system supply (Transmilenio S.A., 2000). This unbalanced growth brought a phenomenon known as "la guerra del centavo," which can be translated as "the war of the cent." The war of the cent refers to the aggressive war of the drivers for picking up the maximum number of potential passengers. This aggressive competition between buses was permitted in the streets. Since the revenue earned by the operators depended on the number of passengers served, the war for passengers was very competitive. Because of the excessive number of buses, these private operators had excessive consumption, tires, and other operational requirements. In addition, the lack of maintenance and renovation of vehicles brought excessive operational costs and increased contaminants and noise. Fare collection was performed by the driver, which produced distractions that in many cases ended in accidents. In addition, this increased the travel time making the service less attractive to the public. Other problems included high pollution levels of 750,000 tons of atmospheric pollutants per year generated by traffic and noise levels above 90dB on major streets. Air pollution was a serious issue due to the higher altitude (27% less available oxygen than at sea level) and the lack of pollution control. In addition, a high number of accidents (52,764) and 1,174 fatalities were recorded in 1998 (Transmilenio S.A., 2001).

4.3.2.2 After the BRT implementation

Transmilenio was created in order to reduce accidents, shorten travel times, reduce pollution, and provide accessibility for young, elderly, and people with disabilities and to provide affordable, high quality and advanced transportation technology. The infrastructure, management, controls and planning are supplied by a new transit authority. The fare collection and operation systems are controlled by the private sector. The new transit authority, Transmilenio S.A. was created in October 1999 in order to manage, control and plan the system. Transmilenio S.A. is supported by 3% of the fare revenues and other activities, such as commercial advertising (Hidalgo and Sandoval, 2001). Financial resources for the implementation of the BRT system came from a fuel tax, local revenues, a credit from the World Bank and grants by the national government. Resources were planned to fund the BRT infrastructure until 2006 with a possible extension to 2018. The project was planned, designed and constructed by local and international firms. It took about eighteen months to finish the studies and develop detailed plans for the system. Examples of BRT systems in other Latin Countries, such as Quito (Ecuador), Curitiba, Sao Paulo, and Goiania (Brazil), and Santiago (Chile) helped to identify important elements for the planning and design of the system.

From the beginning of the BRT implementation the private transportation operators that provided transit service in Bogotá were involved in the planning process. Operators of the old system were offered the opportunity to be the operators of the new system. This strategy was implemented by showing them the opportunities and advantages of their participation. The operators' experience was recognized and valued as a key aspect for the success of the new BRT system. Having the operators of the system as part of Transmilenio, protests and work stoppage possibilities for the service were avoided. Every time that a new Transmilenio bus was put in service, some old buses had to leave the system. The newer buses are used as feeder buses to take passenger from remote locations to the Transmilenio system.

In April 2000, four different firms created by local transportation operators associated with international investors received the contract concession to provide and operate 470 new articulated buses. Ninety-six percent of the private operators that provided transit service acquired stock in the four firms that were awarded the contracts. This shows the success of the program to include former transit operators in the Transmilenio operation. The fare collection was awarded to a local firm supported by an experienced fare collection system provider. The control system was awarded to a Spanish firm. Feeder service and renovation of existing buses contracts were awarded to traditional transit operators (Hidalgo and Sandoval, 2001). The new system infrastructure was constructed by local contractors under the supervision of the Institute of Urban Development (Instituto de Desarrollo Urbano, IDU). Their duties were to develop: 35 km of bus ways and complementary lanes, 4 terminals, 4 parking and maintenance yards, 58 stations, 17 pedestrian overpasses, plazas, sidewalks, built or rehabilitated 126 km of roads for feeder services, in a 24 month construction period. About 17,000 people are estimated to have participated in the project. On December 18, 2000, Transmilenio started operation (Transmilenio).

4.3.2.3 Transmilenio infrastructure characteristics

The Transmilenio infrastructure consists of dedicated bus ways, streets for feeder buses pedestrian access facilities, stations, points for bus parking and maintenance, and an advanced control system. Bus ways are located in the center lanes of the main avenues of the city. These bus ways are physically isolated from the mixed traffic lanes, private vehicles, trucks, and taxis. There are two lanes dedicated for Transmilenio in each direction (See Figure 4. 7). The two lanes in each direction were included to allow buses to pass one another, which improves the speed of the system and allows for express or skip stop service. In a 15 year period, 22 bus ways or main lines covering 388 km are expected to be in operation. Table 4.1 shows the year and the number of km of bus ways that are expected until 2018 (International Seminar on Human Mobility, 2003). Figure 4.8 shows a map of Bogotá with the 22 bus ways expected through 2018.

Year	Km of bus ways
2005	130.4
2010	252.6
2015	384.3
2018	388.9

Table 4.1. Km of Bus ways expected.

Fig. 4.7. Infrastructure Characteristics.

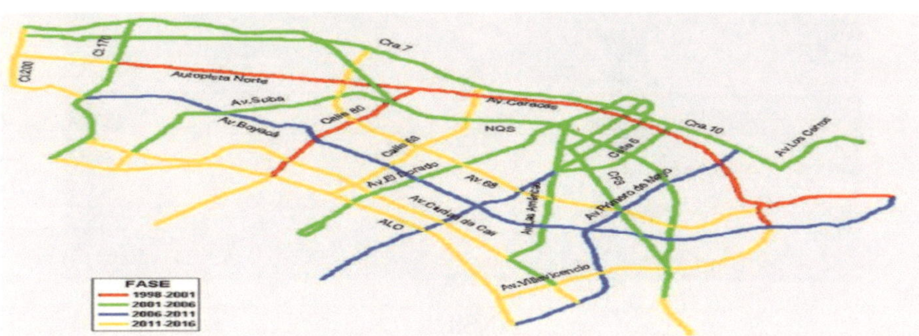

Fig. 4.8. Twenty two Bus ways Through the City.

There are three types of stations;

- Simple Stations: They are located every 500 meters. At these stations passengers can purchase tickets and enter the system.
- Intermediate Stations: These stations are the contact points between the feeder buses and the main lines. Their objective is to provide smooth, fast and effective interaction between the Transmilenio and feeder buses.
- Portals or Main Line Stations: They are located at the beginning and end points of the main line routes. In these stations transfers are accomplished among Transmilenio buses, feeder and transportation routes. The fee is integrated with the feeders, so that when a transfer takes place double payment is not required. These stations are provided with bicycle parking facilities.

OUT SIDE

INSIDE

Fig. 4.9. Stations.

Walkways, plazas and sidewalks were constructed to provide adequate pedestrian and bicycle access. Parking and maintenance areas for the buses near terminal stations were also constructed. Each station is provided with maps and route information to facilitate the use of the system.

4.3.2.4 Operational system

The overall system includes main lines and feeder buses. The buses are an important element for the image of the system. The buses are operated by private contractors, but controlled by Transmilenio. S.A. The main line circulates though exclusive corridors, starting and ending the routes at the Portals and Front End Stations (See Figure 4.9). On the main lines, Transmilenio is the only system operator. Feeder buses do not use the main lines. There are two types of service on the main lines: normal and express service. Normal service stops at every station along the routes; they are identified with the number 1. Express service does not stop at all the stations along the route, which reduces travel time and the size of the fleet because buses can complete more cycles. Express service is identified with the numbers 10, 20, 30, 40, and 50. The combination of the normal and express services allows the system to carry more passengers per hour per direction and divide the passengers according to their destination, which is more appropriate for the size of the buses. Normal buses run every five (5) minutes and express buses every four (4) minutes. Passengers that use the express service can stop and take buses in the other directions as needed. Table 4.2 shows the year and the number of passengers that are expected through 2018 (International Seminar on Human Mobility, 2003).

Year	Passengers/day
2005	2,681,000.00
2010	4,136,000.00
2015	5,004,000.00
2018	5,295,000.00

Table 4.2. Number of passengers expected.

The main line is served by articulated buses with a capacity of 160 passengers. They are 18 meters long, 2.60 meters wide, and have four doors of 1.20 meters each on the left side of the bus. The design of the buses was focused on customers, with inside comfort, easy entrance of passengers, clean air and noise emissions. They have pneumatic suspension, automatic transmission, and the engines use diesel and natural gas. Feeder buses have capacities of 80 passengers (Transmilenio). A total of 6,000 articulated buses are expected by 2015. Figure 4.10 shows a photograph of the buses that serve the feeder system. Feeder buses have lower capacity than the main line buses. New or recent model buses are used as feeder buses with a capacity of 80 passengers. They serve areas that do not have access to the main lines. Passengers transfer from/to the feeder buses to/from Transmilenio though the stations. Each driver works 6hour shifts. Drivers are paid as a function of the kilometers served by their buses, and they are not involved in fare collection. The system operates from 5:00 AM to 11:00 PM. The system is designed to serve 5,000 trips per day with a total investment of US$2.3 billion. This value does not include the fare collection system implementation costs and the cost of the buses

Articulated Buses

Feeder Buses

Fig. 4.10. Articulated buses and Feeder buses.

4.3.2.5 Fare collection system

Transmilenio uses a prepaid method of payment (off board fare collection). The passenger pays the fare upon entry to the system in the stations. The passenger purchases a smart card at the ticket office located at the entrance of each station. The smart card can be yellow for one trip, red for two trips, or the capital card that permits several trips. It permits multiple boarding reducing dwell times, bus operating cost, and travel times for passengers. Access turnstiles are located at the entrances and exits of the stations to validate and register the number of passengers using the system.

4.3.2.6 Advanced control system

An advanced control system is a very important part of the BRT system. A satellite control center allows continue supervision of the operation of the buses. Each bus has a Global Positioning System (GPS) receiver to report the bus location, a computer that contains the schedule, a tracking communication system that shares information with a control center located in Transmilenio, S.A. and the police, and a transponder that sends the information to receivers at the entrances and exits of every station. This communication system provides real time information and is the basis for the control of the system. This makes it possible to adjust the schedule and identify possible new routes into the system.

4.3.2.7 Infrastructure result

Phase 1 was implemented between 1999 and 2002. The system began with 21 stations and was 15.5 km in length. In November 1999 there were 35 km of bus ways, 100 km of feeder routes, 401 articulated buses and 103 feeder buses (Hidalgo and Sandoval, 2001). Three main lines were constructed within the city: Calle 80, Troncal Caracas, and Autopista Norte. Forty-one kilometers of new bus ways were built through the three corridors (See Figure 4.11- This figure also includes phase II). Seven feeder zones with 309 kilometers of feeder routes within 74 neighborhoods were installed to move passengers from remote areas to the main BRT system. Along the main lines four terminal stations, four intermediate integration stations, and 53 simple stations were constructed. In addition, 30 pedestrian overpasses, plazas and sidewalks were constructed. In this phase there were US$240 million of investment (International Seminar on Human Mobility, 2003).

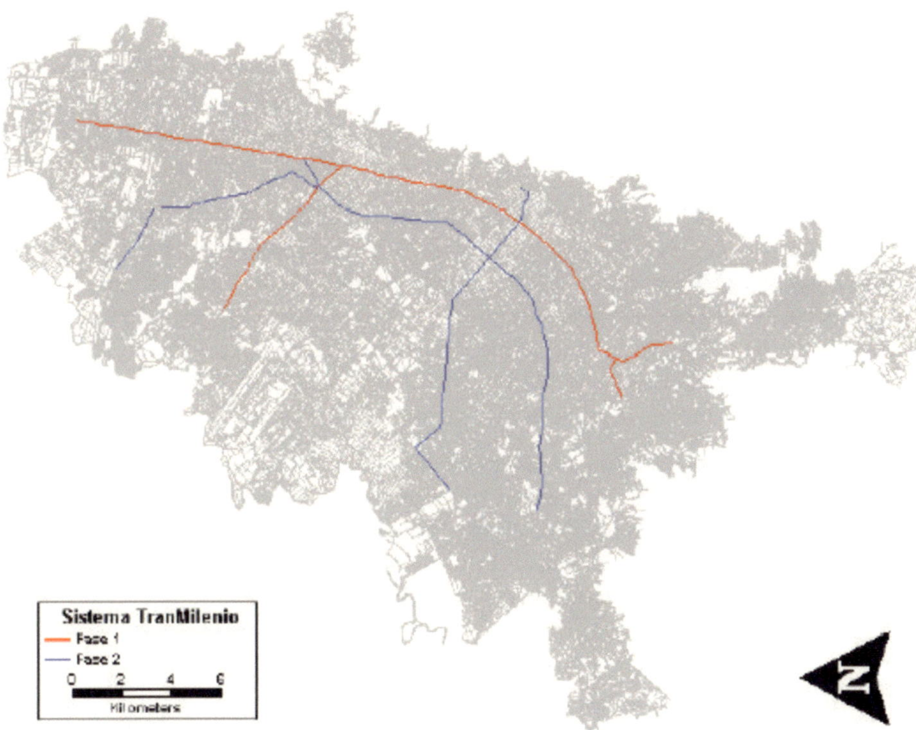

Fig. 4.11. Phase I and Phase II Main Lines.

4.3.2.8 Operational result

In December 18, 2001, the mayor of Bogotá, Enrique Peñalosa, along with Transmilenio S.A. Company, inaugurated the new BRT service in the city. Until January 2001 the service was free. In November, 2000, the system demand grew by 550,000 passengers per week (Hidalgo and Sandoval, 2001). During phase 1 the system moved an average of 770,000 passengers/day with 34,000 passengers per hour per direction on the busy sections of the system, averaging 5.3 passengers per kilometer. A total of 344,162,256 people were transported, and buses covered 66,035,715 kilometers through the main lines (International Seminar on Human Mobility, 2003). There are 470 articulated and 235 feeder buses in use. They have an average speed of 26 km/hour in the main lines. The speed increased from 10 km/hour to 26 km/hour with the implementation of the system (International Seminar on Human Mobility, 2003).

Ticketing and fare collection: In phase I, 90 ticket booths were installed, 359 barriers and 1.3 million smart cards have been used. The fare began at 800 Colombian pesos and ended at 900 Colombian pesos, which is about US$0.40. This low fare makes the system affordable for low income users. This phase had daily revenue of about US$270,000 from about 770,000 passengers (International Seminar on Human Mobility, 2003).

Advanced control system: Six control stations were implemented. Each station is able to monitor and control 80 articulated buses. Each articulated bus has a GPS system to track its location at six-second intervals and with +/-2 meter accuracy. Schedule adherence can be verified and adjusted accordingly. A total of 94 supervisors controlled the buses by the end of phase 1. Continuous communication between operators and the control center supervisors was achieved by the end of this phase (International Seminar on Human Mobility, 2003).

Accident, pollution and safety: The reductions in pollution and accidents as well as safety improvements were some of the most important impacts observed at the end of phase 1. There was observed a reduction of about 92% in fatalities, about 75% in injuries and about 79% in collisions. Robberies at transit stops were reduced by 47% (Hidalgo and Sandoval, 2001). A monitoring study at the one of the main lines (Troncal Caracas) in 2000 and 2001 showed a reduction of about 43% of sulfur dioxide (SO_2), 18% of Nitrogen Dioxide (NO_2), and 12% of particulate matter of less than 10 microns (PM-10) (Hidalgo and Sandoval, 2001).

Travel time: A reduction of 32% in the travel time for public transportation users was measured. The speed along the Calle 80 and Caracas main lines increased from 10 km/Hour and 18 km/hour to an average of 27 km/hour. Surveys show that 83% of the users perceive the increase in speed as the main reason to use Transmilenio. Thirty seven percent of the users perceive that they spend more time with their families because of a faster commute (Hidalgo and Sandoval, 2001).

Hence we can conclude that the various projects in Latin American countries indicate that obstacles have surfaced, but the BRT system is a good alternative to improve and upgrade the transportation system at a cost that developing countries can afford. Examples such as the BRT system in Bogotá, Colombia, demonstrate that the BRT system can be as efficient, cost effective, comfortable and versatile as the LRT. During the first phase of operation the transportation system in Bogotá became more organized and effective and with higher quality. There has been observed a reduction in travel time, pollution, and accidents as well as increases in safety and speed through the network when influenced by the BRT system. People are leaving their cars at home and users seem to accept and like the system. Surveys in Colombia show that the 49% of the users find the system very good and another 49% find the system good during the first phase of operation. The BRT system in Colombia can be taken as an example for other developing countries to follow in the future.

4.3.3 Lesson from Curitiba and Bogota

The example of Curitiba - Brazil and the experience of the Transmilenio BRT system in Bogota - Colombia illustrate the potential of improved bus services to address mobility needs in metropolitan areas. Buses provide flexible and cost effective public transportation. Metropolitan areas throughout the city can build on the experience of Curitiba and other developing cities to develop Bus Rapid Transit systems that provide fast, reliable, and convenient service in cities and suburbs.

Upgrading the performance of bus services to meet the objectives of Bus Rapid Transit will require policies that give priority to bus operations and provide for investment in crucial system components: infrastructure that separates bus operations from general purpose traffic; facilities that provide for increased comfort and system visibility; and technology that provides for faster and more reliable operations. New guidance, information, and fare technologies offer an expanded range of possibilities for operating bus systems that have the potential to produce marked improvements in performance, surpassing previous standards and changing public perceptions of bus service. High quality bus operations have the potential to create new, improved land use options that provide for compact, pedestrian-friendly and environmentally sensitive development patterns that preserve neighborhoods and open space. Bus Rapid Transit thus will have maximum benefit when developed in close coordination with land use policies and community development plans.

Implementation of Bus Rapid Transit poses a number of challenges, ranging from the need for adequate cross sections on city streets to provide separate rights of way for buses, to maintaining the quality of general purpose traffic flow and minimizing local noise and air quality impacts. These challenges require detailed analysis in the context of specific local applications to identify appropriate solutions and to determine where Bus Rapid Transit can have the greatest benefit. Bus Rapid Transit is a concept that merits widespread evaluation and consideration as an adaptable, effective public transportation alternative to automobiles

4.3.4 The bus rapid transit system in Dar es Salaam City – DART system

To mitigate the aforementioned public transport challenges, a Bus Rapid Transit system - so called DART, for Dar Rapid Transit has been proposed since 2003 and succeeded in gaining sufficient funding and political will to be fully implemented by 2030 (JICA, 2008). It is a citywide 137 km system designed to completely replace the existing public transport called daladalas (Dar es Salaam City Council, 2007) Figure 4.12 and 4.13 below is phase 1 and complete route map respectively. It is therefore expected to provide reliable and comfortable

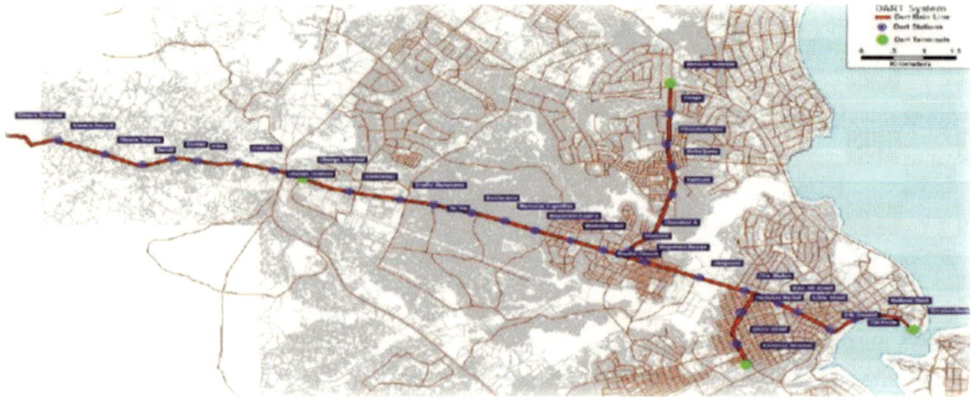

Fig. 4.12. DART Phase 1.

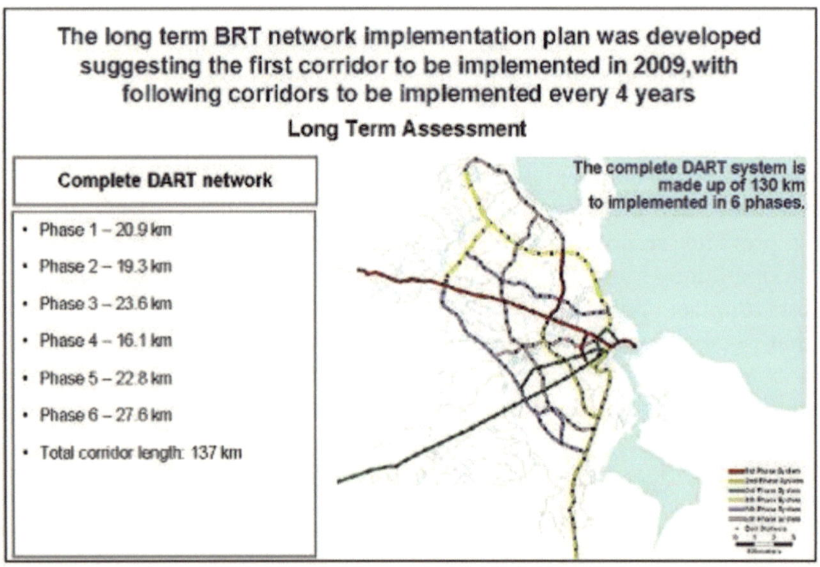

Fig. 4.13. Completed Route Map of DART.

trips with short travel times as its buses run on exclusive and segregated lanes. It is also expected to be more environmentally friendly (and sustainable) as its buses will be EURO 3 standard and thus less polluting than the existing public transport (daladalas). Although the following specific questions have been developed: Would the BRT buses ensure equitable access to all population groups in the city? Does the new system accommodate the current public transport operators? How does vehicular emissions of CO2 and CO in the city before and after Dart implementation.

4.4 Evolution of road network and BRT coverage in Dar es Salaam City

A total of 1,091 km roads including the proposed expressway are expected at the year of 2030 (Figure 4.14 and 4.15) respectively. Most of the roads in Kigamboni side are new roads that are expected to attract future urban development in the south of Dar es Salaam. An expressway system is proposed, which runs through the entire Urban Growth Boundary (UGB) area as a spine road system for the region. The transport policy suggests that the priority should be given to people's mobility, but at the same time mobility of cars is important in order to improve the attractiveness of Dar es Salaam as a Gamma World City in future. The expressway runs in parallel with the Morogoro road in the future urban business/commercial axis, hence which provides direct access with motorists to go to their destinations along the BRT Phase 1 corridor. This expressway alignment will contribute to the urban regeneration of the BRT Phase 1 corridor as well

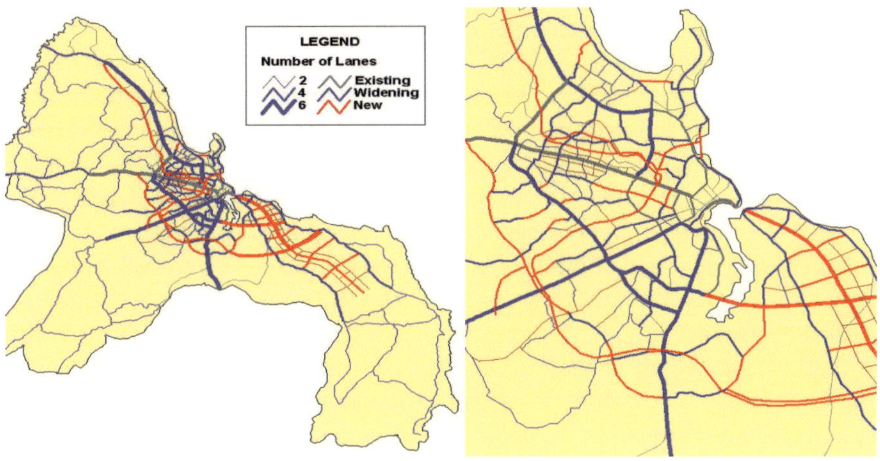

Fig. 4.14. Year 2030 Network by the number of lane.

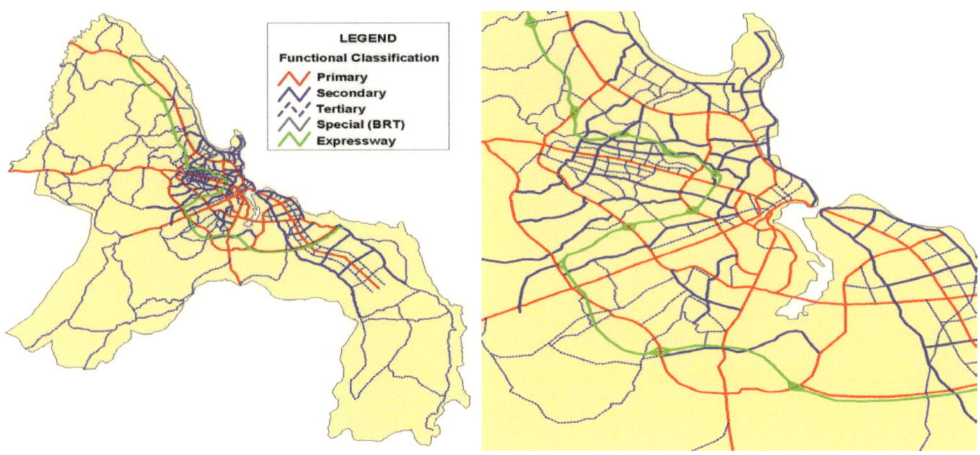

Fig. 4.15. Year 2030 Network by road class.

4.4.1 Network planning and efficiency

Dar es Salaam is well progressed to developing a 'state of the art' BRT system but it is necessary to place this into a proper and objective planning context. BRT is not a 'cure all' and its success is dependent on where it sits in the overall public transport of Dar es Salaam and whether it meets the essential criterion necessary for success. All successful public transport systems need to include the demand oriented services; Bus priority (including BRT where appropriate) giving; increased bus speeds; reduced travel time; access and integration; efficient network design.

4.4.2 Demand oriented services

Under traditional public ownership, buses were mostly 'supply oriented' building the system on technical or regulatory premises on the assumption that patronage will follow. More recently, transport planners have understood that consumers have greater choice, so for public transport to survive demands a 'demand responsive' approach. For the transit user, the transit experience is more than just a bus trip; it is a total journey (from door to door) and judged in terms of access, convenience, travel time and comfort. Network design should take this into account and involve all aspects of route planning, passenger waiting facilities, ticketing, bus quality, service frequency, comfort and convenience in the planning process. For a 'demand oriented system', passenger convenience is the main issue, being that the system can be used with ease and ultimately saves the user time. This concept needs to be embedded into policy objectives and the design and management of the system. The figure: 4.16 and figure: 4.17 below show the bus priority and the role of DART manager respectively. Each city must develop its own strategic approach for Bus Priority and BRT as 24 part of the total system network. BRT is not an 'off the shelf' bus solution; it needs to be carefully adapted to the prevailing conditions, but in Dar es Salaam, high public transport dependency creates a natural opportunity for a BRT system and hence to insure the role of Dart manager in the implementation and running the system management model.

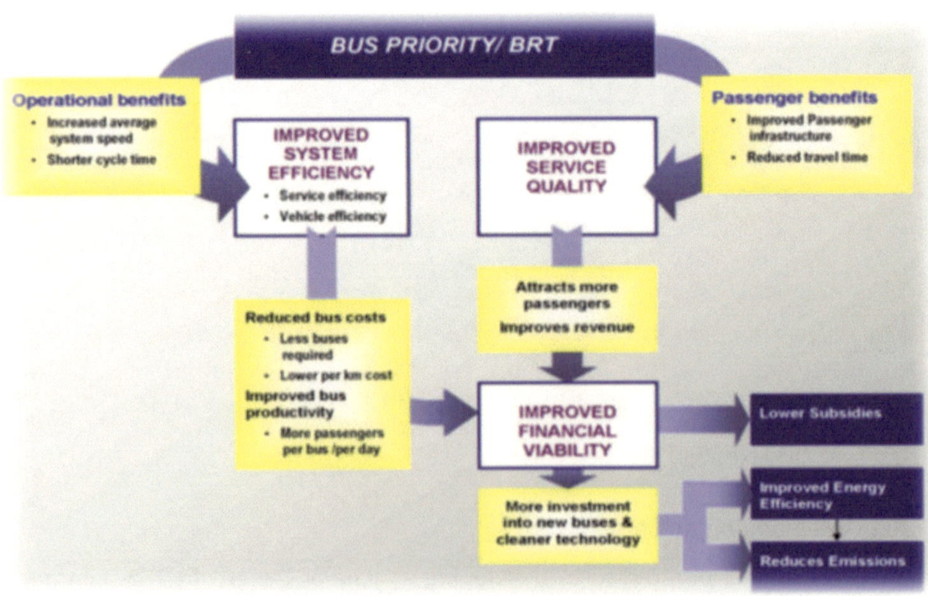

Fig. 4.16. Bus priority.

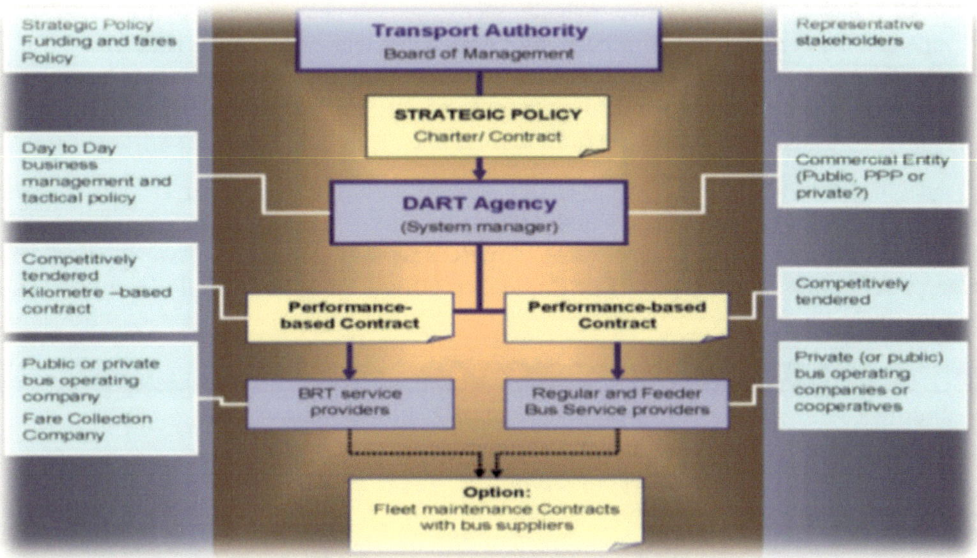

Fig. 4.17. System Management Mode.

4.4.3 DART system

Dar Rapid Transit - DART will be a high quality mass transport system in Dar se Salaam - Tanzanian operating on specialized infrastructure and offering affordable mobility, environment improvements , and a batter quality of life of the resident of Dar es Salaam. The DART mission is to provide quality and affordable mass transport system for the residents of Dar es Salaam which will reduce emissions, enable poverty reduction, lead to sustainable economic growth, improve the standard of living and act as a pioneer of private and public investment partnership in the City. DART will achieve this by using: Modern, privately managed buses with low emissions; segregated bus lanes; Scheduled bus services; High capacity bus stations with a central platform; on level boarding; Privately-managed fare collection system; and Average bus speeds over 22 km/hr.

The system has five system actors. The DART Agency, a public agency, will regulate and manage the system and the service. Two private sector bus companies and one private sector fare Collector Company will provide the system services, which will be publicly bid. The winners of the bid will be awarded 20 year concessions. Finally, one fund manager will ensure fiscal accountability and transparency of the fare revenue.

i. *DART Agency* - a government agency responsible for overall management of the system, policy--setting, regulation, planning and controlling of operations and marketing.
ii. *DART Fare Collector* - one private sector company responsible for daily fare collection, maintenance of the physical infrastructure at the bus stations, acquisition and maintenance of equipment used at the station, as well as cleanliness and security at stations.

iii. *DART Bus Operator(s)* - two private sector companies responsible for acquisition, operation and maintenance of buses along specified trunk and feeder routes.
iv. *DART Fund Manager* - one institution responsible for financial management and reporting, liquidity control and payments to the system actors (bus operators, far collector, DART agency and the fund manager).

All system actors will adhere to the DART system values of: Customer Driven and User Friendly, Innovative, Cost Effective and Affordable, Timely and Reliable, Team Work, Safe, Financially Sustainable and Profitable, Environmentally Friendly, Poverty Reduction through Economic Development. Dar Rapid Transit will provide several direct benefits to the City, including:

Category	Description
Economic	Cheaper means of mass transportation Reduce travel times Less congestion Segregated lanes for buses Fast boarding and disembarkation Centrally controlled communications system Increase economic productivity fair return on investment building of Public Private Partnerships Less cost for customers and Government in general Better working conditions for drivers and conductors Create employment
Social	Reliable and comfortable Reduce accidents and air pollution-related illnesses Increase civic pride and sense of community
Environmental	Less vehicle pollutants emissions Reduce noise levels Prioritizes and integrates Non Motorized Transport
Political	Delivery of mass transit system within one political term Delivery of high-quality resource that will produce positive results for virtually all voting groups
Urban form	More sustainable urban form City Beautification

Table 4.3. Dar Rapid Transit benefits.

4.4.4 DART will be the first full BRT system in Africa

DART is a high capacity bus rapid transit system with closed stations and physically segregated lanes. Phase 1 of the system consists of 21 km of segregated median bus ways, 29 stations, 6 feeder stations, 5 terminals, and a network of feeder routes operating in mixed traffic. It will offer seven trunk line services, using (136) air conditioned, 18 meter articulated buses with a 140 passenger capacity and 15 feeder bus services, using (111) air conditioned, 8.5 meter micro - buses with a 50 passenger capacity. The baseline proposed fare to the public will be: Table 4.4 below highlight the proposed fare.

Trunk and Feeder	500Tshs
Trunk Only	400Tshs.
Feeder Only	400Tshs.

Table 4.4. Proposed fare highlights.

Customers will enter stations and feeder buses with contactless smart cards. There will be an option to pay cash both on the feeder buses and in the stations. The DART Agency has set baseline fares at Tsh 400 for trunk service and Tsh 400 for feeder service. A customer transferring from a trunk bus to a feeder or vice versa will pay a combined fare of Tsh 500 if using a contactless SMART card. Those using cash must pay Tsh 400 to use the trunk bus services and Tsh 400 to use the feeder services. There will be 392,217 passengers per day on the system for a total annual ridership of 117.7 millions of trips. The projected total fare income is estimated to be TZS 46 billion in the project's first year of operation, rising to TZS 73.6 billion in 2028. Table 4.3 below highlights the system's revenue stream for the first 20 years of operations.

Revenue Allocation ('000 TZS)	2009	2010	2011	2012	2013
Tariff revenue	46,162,583	47,752,845	49,397,890	51,099,605	52,166,113

Table 4.5. System's revenue highlights.

The goal of DART is to transform Dar es Salaam's informal system of "daladala" minibuses into a customer--oriented public transportation service. To this end, the DART Agency will prohibit 43 of the daladala routes that utilize Morogoro and Kawawa North Roads. The Agency plans to cancel all of the daladala routes on Morogoro Road. On Kawawa North Road, daladalas continuing south, away from the BRT corridors, will be permitted to operate, but those turning onto the Morogoro corridor will be cancelled. Table 4.4 below is the operation highlights of the DART system.

	Trunk	Feeder
Peak headway (min.)	1.5	1.1 to 7.5
Off--peak headway (min.)	6	15
Average commercial speed (km/hr)	23	17
Hours of operation	5:00 to 23:00	5:00 to 23:00
Daily service kilometers	45,000	26,000

Table 4.6. Operational highlights.

4.4.5 System scheduling and control

There are two parts to the system control. The first is scheduling the services - every week, every day, and every hour for each service (all initial 7 trunk services and 15 feeder route services) for each operator. The second is controlling the system to ensure quality of the service, including if the buses are running on time and know if the drivers are going too slow or too fast. Control also involves addressing contingencies if there are problems in operations (i.e. a bus breaking down, strikes) and the ability to speak directly with drivers to let them know about all these conditions. In order to ensure system cohesion and technical integration, as well as achieve economies of scale, the FCS operator is responsible for providing:

The scheduling software and the control software to the DART Agency, training on the respective software's to the DART Agency; the communications and physical equipment for the control functions of the system, including data transmission, on board logic units and GPS on the buses, and radio services; terminals at the bus depots with software for receiving the scheduling from the DART Agency; WLAN points at the depots to facilitate data transmittal; and Specifications to the bus operators for the Logic Units that are on--board the buses to do the control so that they can plan for them in the bus layout.

i. *Scheduling* - The DART Agency schedules the bus service. The two main inputs are the number of buses (which is a fixed number) and the number of total passengers and number per station (which is dynamic and comes from the Fare Collection System information). While the bus operators are responsible for providing the scheduled service, the FCS operator is responsible for providing the number of passengers. Since it is the FCS operator's responsibility to provide the passenger number information, it is also responsible for providing the software for scheduling to the DART Agency to ensure coherence between the information from the FCS and the software used to do hourly, daily, weekly, monthly scheduling per service. The DART Agency will create monthly, weekly, and daily schedules for the bus operators. They produce the longer term schedules in order to help the bus operators schedule maintenance for the buses.

The DART Agency sends to the depot of each operator the daily schedule (routes and services) at least one day in advance for all services, including truck and feeder. There will be one terminal (with appropriate software) in each depot provided by the fare collector that will receive that information. The bus operator then takes that information and assigns specific buses and specific drivers to the service. The software the bus operator uses to do that bus allocation and crew scheduling is their own responsibility. The bus operator sends the scheduling with the specified bus and driver back to the scheduler in the DART Agency control center. The control center then must verify that it matches the schedule that was sent originally and do their first control looking at what buses are running to ensure buses are being rotated for maintenance and how long the drivers are scheduled to work to ensure that safety and legal standards are met The control center then must approve the schedule. Once the bus operator gets the approval, they then must save the information of the service of each bus on the Logic Unit (on board computer) of each bus, both trunk and feeder. The Logic Unit is bought by the Fare Collector. However the technical specifications must be given by the Fare Collector to the bus operators before they procure the buses in order for the manufacturers to know the cable specifications, as well as support required for and the dimensions of the Logic Unit. When the bus operators send information of detailed scheduling back to the control center, they also save it on the on board computer to help the driver be on time. Station stops and times appear on the logic unit's display for the driver. Only the driver assigned to a particular bus may operate that vehicle. This can be ensured by giving each driver a specific code that unlocks the bus. For a particular bus, both the service and the driver will change daily. The important consideration is to save the correct service and driver information to the bus specified for that day's service. For feeder buses, the driver is joined by a conductor working for the Fare Collection Company. The bus operator and the Fare Collection Company will have to work together to coordinate the schedules of personnel (conductors on the buses) and coordinate the timing issues with depositing the money collected each round trip. Finally, the card readers on the feeder buses are the responsibility of the Fare Collection Company to maintain, but there will be fines to the bus operator for negligence if equipment is damaged. The bus operator is not paid for dead kilometers meaning the kilometers from the bus depot to the start of the service and the kilometers from the end of the service to the depot. They are only paid for the kilometers of service provided. Once the bus is in service, the driver only talks to and follows directions from the DART control center.

ii. *Controlling* - The main form of control will be using GPS tracking and sending that information to the control centre via GPRS data packets. The buses will be equipped with Global Positioning System (GPS) transponders that are procured by the Fare Collector in the Logic Unit. These transponders will record arrival times and location of the bus and periodically during the day this information will be sent to the control center for verification. The distance traveled will be determined by the control center computer and checked against odometer readings, beginning with the place where service was started and ending where the service finished. With GPS, the central computer system will record arrival times at stations and terminals and then check them against the assigned schedule. The bus driver proceeds to drive the scheduled service that is on the Logic Unit. The bus driver is charged with maintaining the service and the frequency on the schedule and adjusting his driving to match those specifications.

However, this will also be monitored by the control center that can then request that the driver to slow down or speed up to meet the assigned schedule. While on the road, bus drivers have direct contact with the control center. This communication can be through radios or using cell phone technology / GSM chips in the logic unit with pre configured panel activation buttons. As soon as the driver is on the corridor, he should only take direction from one source and that source should be the DART Agency control center never from the bus operator office or depot. The logic unit will also record the number of kilometers driven and the times the driver arrived at the station or predefined Feeder Bus Stops, as determined by the GPS locator and recorded by the Logic Unit. This information will be sent to the control center via GPRS technology at pre determined intervals. There will be a mechanism on the bus that the driver activates in order to open the sliding doors at the station. When the bus returns to the depot, all that information from the Logic Unit, including the odometer reading, will be sent to the control center by the Depot Terminal. For manual control, when the bus reaches the first terminal, the DART Agency controller visually verifies the time of arrival and the condition of the bus. If there are any problems, the controller can contact the control center to remedy via radios or GSM chip / cell phones. If there is a problem in the system (bus early, bus late, bus in poor condition), the people in the terminals must transmit immediately to the control center for them to fix directly with the driver or with the depot, in order to send a reserve bus, for example. The timing information will also be transmitted via GPRS from the bus directly. The control center is responsible for comparing scheduling versus actual, both the time scheduling and the kilometers. Those numbers get shared with the financial planning team. The onboard computer and software is provided by the FC as a bundled unit to the bus operators and the FC is responsible for ensuring / guaranteeing data transmission. Figure 4.18 below illustrates the control system.

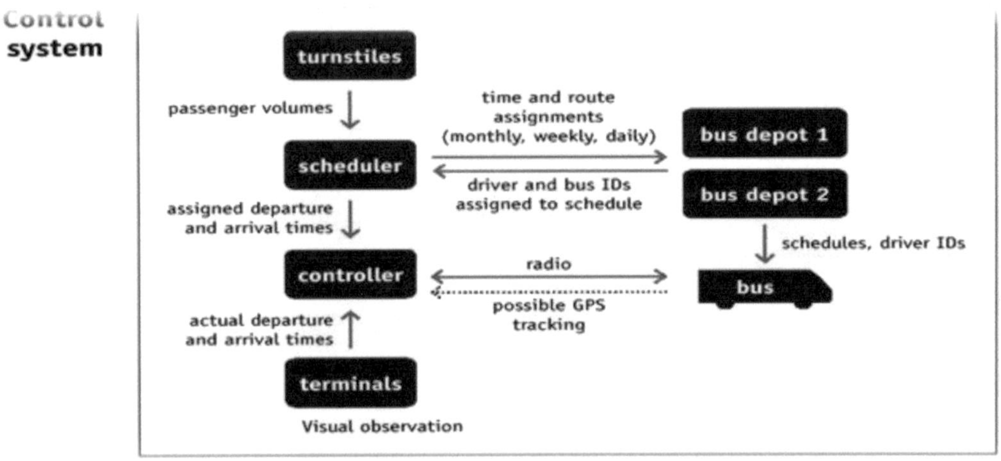

Fig. 4.18. Control system.

4.4.6 Fund manager

The transparent and fair distribution of revenues is fundamental to the DART System and in operating a network of integrated transit providers. (Figure 4.19 the passenger revenues). To ensure confidence in the distribution of revenues in the private sector, as well as the DART Agency, an independent trust fund manager acts as a custodian of the revenues and pays out according to fixed rules and regulations as stipulated in the contracts. This fund manager will be paid a fixed fee for maintaining this fund and paying the appropriate operators. This fund manger receives the money every day from operations (from the fares) and then pays all the system operators based upon the contracts and the information given to them by the DART Agency. The fund manager will be responsible to the DART Agency and will carry out the following functions:

i. Put systems in place for the management of the DART resources;
ii. Makes payments to the various actors within the DART system upon instruction from the DA. This will be on the basis of pre agreed terms;
iii. Prepare regular financial reports for submission to the DA.

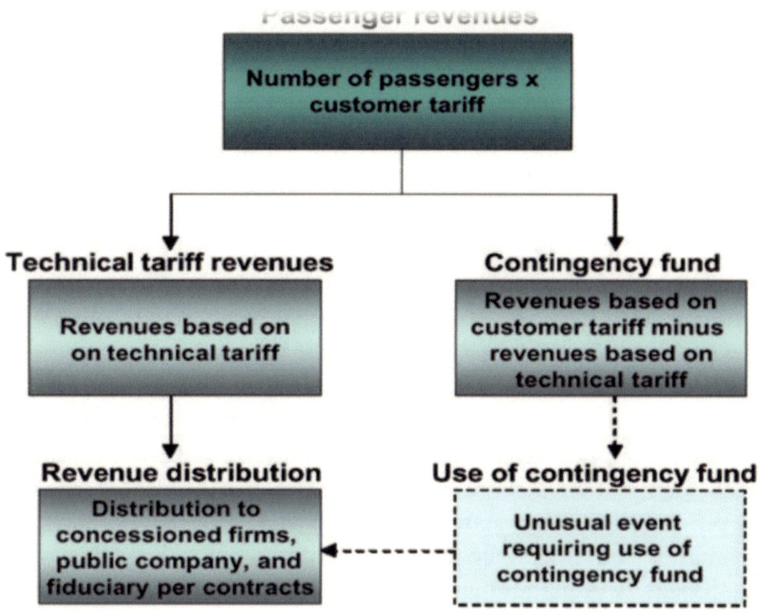

Fig. 4.19. Passenger revenues.

The fare collection system is managed by a separate private company that successfully bid for the fare handling concession. The fare handling company has no involvement with any of the bus operating companies on the BRT system. Since the fare collection company itself is due part of the proceeds, it would be a source of potential suspicion if the fare collection company was to fulfill the function of fund manager. This ticketing system operator collects the fare revenue and deposits it into the account of the trust fund manager. Finally, to ensure transparency of the system, the entire process will be independently audited by another professional firm. This auditing process provides a check on the handling of revenues by the fare collection company and the fiduciary company. The auditing process in conjunction with the electronic verification of fares collected, as well as the presence of the fiduciary company, all help contribute to an environment of confidence in the system. The figure 4.20 below shows how the revenue will be managed in the DART System.

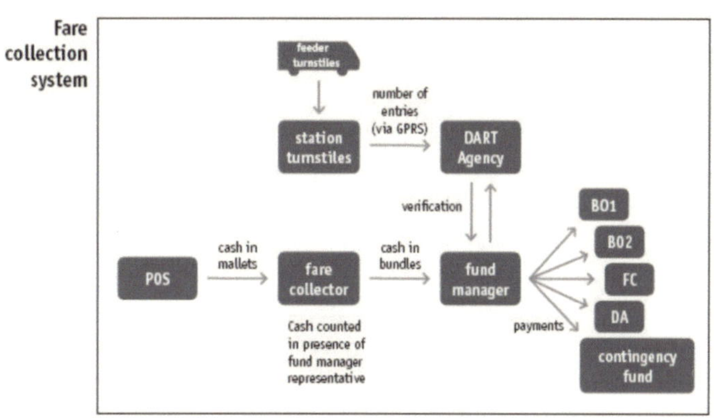

Fig. 4.20. Dart system revenue management.

4.4.7 Dart project risks and mitigating actions

The success of the DART Project is challenged by the following key factors:

i. Price sensitive market. The majority of commuters in Dar es Salaam fall in the low income bracket that is sensitive to price changes. A contingency fund will therefore be established to cushion against sudden or unusual fare increases.
ii. Unstable fuel prices. The operators will strive to form strategic alliances with oil companies and possibly enter into forward contracts that will shield against unforeseen fuel price increases.
iii. High borrowing costs. The project requires specific buses, equipment, computers and technology that are of modern standards. Acquisition of most of the project assets requires external financing. In Tanzania interest rates are high and this can only be minimized by the DART system operators obtaining a syndicate loan(s).
iv. Unreliable power supply. Power fluctuations and interruptions are common in the city and this can disrupt DART services. Power generators have been considered as an alternative power source at bus stations and terminals.
v. General resistance from the public (commuters) and other stakeholders. An aggressive marketing strategy is in place to enable the DART Agency to sensitize the public and create awareness or acceptance.
vi. Security of cash in transit. The risk of theft or robbery of DART collections during transit points i.e. to and from stations/terminals and the bank can be mitigated by enlisting the services of a reputable and experienced bank offering secure cash-in transit services.

Hence in order to achieve the above revenues targets, the Project will consider the following critical success factors: Tax waivers on bus importation; banning of dala dalas (and enforcement of ban) from Phase One Corridor; obtaining financing for the project at an affordable rate; close collaboration of the public and private sectors.

5. Intelligent transportation system technology in bus rapid transit

The use of new Intelligent Transportation Systems (ITS) or Advanced Public Transportation Systems (APTS) applications could contribute to improved bus service and increased bus operating speeds. Some ITS and APTS applications that a Bus Rapid Transit system might employ are described below, but this list is by no means exhaustive:

i. *"Smart" card fare collection methods* - use read and write technology to store dollar value on a microprocessor chip inside a plastic card. As passengers board a bus, the card reader determines the card's value, debits the appropriate amount for the bus ride, and writes the balance back onto the card, all within a fraction of a second. There are two types of card readers, the proximity reader which can read cards held a few inches away, and the contact reader which requires physical contact with a card. Under development are systems that will be able to read cards carried in passengers' pockets, wallets and purses. Cashless systems such as "smart" cards speed up the fare collection process and eliminate expensive cash handling operations at transit agencies. "Smart" cards can also be programmed for distance - based pricing by recording where a passenger enters a transit system and debiting the appropriate amount from the card balance according to the point where the passenger exits the system, regardless of the number of internal transfers.

ii. *Automatic vehicle location (AVL) systems* - enable transit agencies to track their vehicles in real time and provide them with information for making timely schedule adjustments and equipment substitutions. AVL systems are computer based vehicle tracking systems that measure the actual real time position of each vehicle, and relay the information to a central location. The measurement and relay techniques vary, but the most common are: signpost and odometer, wherein a receiver on a bus detects signals sent by signposts along the bus route and transmits the identity of the signpost and the odometer reading to the control center; and global positioning satellite (GPS) technology, wherein an onboard GPS receiver determines the bus position and transmits the information to the control center. AVL systems can be augmented by geographical information systems (GIS) on control center computers that display the location of the vehicles on route map grids.

iii. *Computer aided dispatching and advanced communications* - are systems that enable transit dispatchers, in combination with AVL systems, to maintain bus system efficiency by performing service restoration activities and communicating instructions to and receiving messages from drivers. Service restoration activities include such operations as adjusting dwell times at bus stops or transfer points, adjusting vehicle headways, rerouting vehicles, adding buses to routes, and dispatching new vehicles to replace disabled vehicles. Communications can be received in buses via radiotelephones, cellular telephones, or mobile display terminals.

iv. *Precision docking at bus stops* - uses sensors on buses and on the roadside to indicate the exact place where the bus should stop. Bus doors opening at the same location each time make it possible for passengers to be in position for immediate boarding once a bus has stopped, shortening dwell time.

v. *Tight terminal guidance* - uses sensors similar to those for precision docking to assist buses in maneuvering in terminals with limited space. This type of system can help minimize the amount of space needed for bus terminal operations, as well as reduce the overall amount of time a bus spends at terminals.

vi. *Warning systems* - are beginning to appear on the market to assist the bus driver in a number of safety areas: collision avoidance, pedestrian proximity warning, attentive driver monitoring and warning, intersection collision avoidance, and low tire friction warning. Safety improvements can help any bus system increase its reliability and efficiency by reducing the likelihood of accidents and incidents.

vii. *Passenger information systems* - give passengers the means to make informed decisions about their transit travel. Of the many technologies now available for passengers to access this type of information, the APTS applications most appropriate for Bus Rapid Transit are in vehicle information systems. These systems automatically announce approaching bus stops, allowing disembarking riders to position themselves near the doors prior to arriving at their stops, and speeding up the unloading and loading operation.

viii. *Automated enforcement systems for exclusive bus lane* - are being enhanced by new technology, including automatic video cameras and infrared sensors. These state-of-the art systems are just now appearing on the commercial market.

6. Effect of bus rapid transit

Successful Bus Rapid Transit systems can be expected to produce improvements in bus service, operations, and ridership, and to affect traffic congestion and air quality:

i. *Bus speeds and schedule adherence* - Perhaps the most fundamental effect of a Bus Rapid Transit system, travel times would likely improve due to the lack of impediments to bus movement along exclusive bus lanes. Bus speeds would be expected to improve not only in absolute terms, but also relative to the automobile traffic that parallels the exclusive lanes.

ii. *Ridership* - Ridership would be expected to increase due to improved bus speeds and schedule adherence. Customers who use buses infrequently might ride more often, and some automobile users might convert to transit. A visible improvement in bus speeds might be noticeable to drivers of other vehicles, presenting a positive image of transit as an alternative to driving.

iii. *Other traffic* - If the creation of exclusive bus lanes reduces the number of lanes available for other traffic, then in the short term the possibility of increased congestion on the roadways is raised. Traffic flow on cross streets and turning traffic may be disrupted as buses use their signal priority to travel uninterrupted through intersections. Further, mobility on alternate routes may deteriorate, as drivers seek ways to avoid roads with exclusive bus lanes. One of the challenges of implementing an exclusive bus lane would be to minimize this disruption.

iv. *Air quality* - Long term, as ridership increases and the overall level of general purpose traffic decreases, urban areas may experience improved air quality due to reduced emissions from automobiles

7. Conclusion

The example of Curitiba - Brazil, the experience in the Bogota – Colombia, South Africa and the application of Bus Rapid Transit system in Dar es Salaam Tanzania illustrate the potential of improved bus services to address mobility needs in metropolitan areas. Buses provide flexible and cost effective public transportation. Metropolitan areas throughout the developing cities can build on the experience of Curitiba, Bogota and other developing cities to develop Bus Rapid Transit systems that provide fast, reliable, and convenient service in cities and suburbs. Upgrading the performance of bus services to meet the objectives of Bus Rapid Transit will require policies that give priority to bus operations and provide for investment in crucial system components: infrastructure that separates bus operations from general purpose traffic; facilities that provide for increased comfort and system visibility; and technology that provides for faster and more reliable operations. New guidance, information, and fare technologies offer an expanded range of possibilities for operating bus systems that have the potential to produce marked improvements in performance, surpassing previous standards and changing public perceptions of bus service. High quality bus operations have the potential to create new, improved land use options that provide for compact, pedestrian friendly and environmentally sensitive development patterns that preserve neighborhoods and open space. Bus Rapid Transit thus will have maximum benefit when developed in close coordination with land use policies and community development plans. Implementation of Bus Rapid Transit poses a number of challenges, ranging from the need for adequate cross sections on city streets to provide separate rights of way for buses, to maintaining the quality of general purpose traffic flow and minimizing local noise and air quality impacts. These challenges require detailed analysis in the context of specific local applications to identify appropriate solutions and to determine where Bus Rapid Transit can have the greatest benefit. Bus Rapid Transit is a concept that merits widespread evaluation and consideration as an adaptable, effective public transportation alternative to automobiles that has the potential to meet a broad range of mobility needs and support an improved quality of life.

8. Acknowledgment

The author would like to express appreciation to Prof. Yan Chen for guidance and Dr. Qu Lili for providing useful contact. This work was supported by specialized research fund for the doctoral programme of Higher Education (No 200801510001) and National Natural Science foundation of China (No. 70940008).

9. References

A. K. Bit, M. P. Biswal and S. S. Alam (1993), Fuzzy programming approach to multi objective solid transportation problem, *Fuzzy Sets and Systems*, Vol. 57, pp. 183–194.

Barrios, S., L. Bertinelli & E., Strobl. (2006). Climate Change and Rural-Urban Migration: The Case of Sub-Saharan Africa. Journal of Urban Economics, 60, 215-218.

Button, K. & P. Nijkamp (1997). Social Change and Sustainable Transport. Journal of Transportation Geography, 5 (3), 215-218.

Bus Rapid Transit, Sustainable transport: A sourcebook for policy-makers in developing cities. Eschborn: Deutsche Gesellschaft fur Technische Zusammenarbeit (GTZ) GmbH.

Dar es Salaam City Council. 2007. Consultancy Services for the Conceptual Design of a Long term Integrated Dar es Salaam BRT System and Detailed Design for the Initial Corridor:

E. L. Hannan (1981), Linear programming with multiple fuzzy goals, *Fuzzy Sets and Systems,* Vol. 6, pp. 235–248.

Faiz, A., C.S. Weaver & M.P. Walsh. (1996). Air Pollution from Motor Vehicles: Standards and Technologies for Controlling Emissions. World Bank.

G. L. Thompson (1997), Network models, in *Advances in sensitivity analysis and parametric programming,* T. Gal. and H. J. Greenberg (eds.), Kluwer, Boston, pp. 7-1–7-34.

H. J. Zimmermann (1997), Fuzzy linear programming, in *Advances in sensitivity analysis and parametric programming,* T. Gal and H. I. Greenberg (eds.), Kluwer, Boston, pp. 15-1–15-40. *Received April, 2005*

Habitat, United Nations Center for Human Settlements. (1996a). An Urbanizing World:Global Report on Human Settlements. Oxford University Press, Oxford.

Hall, F. (2004). Effectiveness of Participatory Planning on Urban Transportation in Dar es Salaam, Tanzania. Proceeding. Cooperation for Urban Mobility in the Developing World (CODATU) 10th Conference. Bucharest, Rumania. Available on http://www.codatu.org/francais/publications/actes/conferences/codatu11/Papers/halla.pdf Accessed on 19 April 2010

Hidalgo, D. and Sandoval, E. (2001). *Transmilenio: A High Capacity – Low Cost Bus Rapid Transit System Developed for Bogotá, Colombia. Transmilenio S.A...* Bogotá, Colombia.

H. Leberling (1981), On finding compromise solutions in multi criteria problems using the fuzzy min-operator, *Fuzzy Sets and Systems*, Vol. 6, pp. 105–118.

JICA, Japan International Cooperation Agency. (2008). Dar es Salaam Transport Policy and System Development Master Plan. Technical Report. Pacific Consultants International.

J. L. Ringeust and D. B. Rinks (1987), Interactive solutions for the linear multiobjective transportation problem, *European Journal of Operational Research*, Vol. 32, pp. 96–106.

J. Current and M. Marish (1993), Multiobjective Transportation network design and routing problems, *European Journal of Operations Research*, Vol. 65, pp. 4–19.

J. A. Diaz (1978), Solving multiobjective transportation problem, *Ekonomicko Matematicky Obzor*, Vol. 14, pp. 267–274.

J.V. Beaverstock, R.G. Smith and P.J. Taylor. *A Roster of World Cities*, in Cities, 16 (6), (1999), pp 445-458

J. N. Clirnaco, C. H. Antunes and M. J. Alves (1993), Interactive decision support for multi-objective transportation problem, *European Journal of Operational Research*, Vol. 65, pp. 58–67..

Kanyama, A., A. Carlsson-Kanyama, A. Linden & J. Lupala. (2004). Public Transport in Dar es Salaam, Tanzania. Institutional Challenges and Opportunities' for a Sustainable

Lupala, J. (2002). Urban Types in Rapidly Urbanizing Cities: Analysis of Formal and Informal Settlements in Dar-es-salaam, Tanzania. Stockholm Royal Institute of Technology.

L. Li and K. K. Lai (2000), A fuzzy approach to the multi objective transportation problem, Computers and *Operations Research*, Vol. 27, pp. 43–57.

L. H.Chen and F. C. Tsai (2001), Fuzzy goal programming with different important and priorities, *European Journal of Operational Research*, Vol. 133, pp. 548–556.

M. K. Luhandjula (1982), Compensatory operators in fuzzy programming with multiple objectives, Fuzzy *Sets and Systems*, Vol. 8, pp. 245–252.

M. Sakawa (1988), An interactive fuzzy satisfying method for multi objective linear fractional programming problems, *Fuzzy Sets and Systems*, Vol. 28, pp. 129–144.

OECD, Organization for Economic Cooperation and Development. (2000). Synthesis Report of the OECD project on Environmentally Sustainable Transport. Presented on the International Conference 4th to 6th October 2000 in Vienna, Austria.

Olvera, L., D. Plat & P. Pochet. (2003). Transportation conditions and access to services in a context of urban sprawl and deregulation: The case of Dar es Salaam. Transport Policy, 10, 287-298.

Olvera, L., D. Plat & P. Pochet. (2008). Household Transport Expenditure in Sub-Saharan African Cities: Measurement and Analysis. Journal of Transport Geography, 16, 1-13.

R. Isermann (1979), The enumeration of all efficient solutions for a linear multiple-objective transportation problems, *Naval Research Logistic Quarterly*, Vol. 26, pp. 123–139.

R. E. Bellman and L. A. Zadeh (1970), Decision-making in a fuzzy environment, *Management Science*, Vol. 17, pp. 141–164.

S. Chanas and D. Kuchta (1998), Fuzzy integer transportation problem, *Fuzzy Sets and Systems*, Vol. 98, pp. 291–298

S. Chanas,W. Kolodziejczyk and A. Machaj (1984), A fuzzy approach to the transportation problem, *Fuzzy Sets and Systems*, Vol. 13, pp. 211–222.

S. Chanas and D. Kuchta (1996), A concept of the optimal solution of the transportation problem with fuzzy cost coefficients, *Fuzzy Sets and Systems*, Vol. 82, pp. 299–305

SSATP, Sub-Saharan Africa Transport Policy. (2005). Non-motorized transport in African cities. Lessons from experience in Kenya and Tanzania. SSATP Working Paper No. 80. World Bank. 35

Sassen S. Cities in a World Economy. Pine Forge Press, London. 1994; Sassen S. The urban complex in a world economy. In International Social Science Journal 139 pp 43-62. 1994.

The United Nation Population Fund. New York. US EPA, United States Environmental Protection Agency. (2005). Emission Facts

UNFPA, The United Nation Population Fund. (2004). State of the World Population 2004. The United Nation Population Fund. New York.

UNFPA, The United Nation Population Fund. (2007). State of the World Population 2007: Unleashing the Potential of Urban Growth. .

Vuchic, Vukan R. 2005. Light rail and BRT: Competitive or complementary? Public Transport International 2005 (5):10-13.

Whitehand, J., & J. Larkham. (1992). Urban Landscapes: International Perspectives. Routledge. Whitelegg, J. (2003). Selected International Transport Investment and Funding Frameworks and Outcomes. Final Report. Australian National Transport Secretariat. Available on:
http://www.eco-logica.co.uk/pdf/australia.pdf Accessed on 20 Mach, 2010.

World Commission on Environment and Development, WCED (1987). Our Common Future.Report of the World Commission on Environment and Development. UN Documents Cooperation Circles. New York.

WUP, World Urbanization Prospects. (2005). World Urbanization Prospects: The 2005 Revision.Executive Summary Fact Sheets Data Tables. United Nations, New York. Yin, R. (2003). Case Study Research: Design and Methods. Sage. London. 36

Wright, Lloyd. 2001. Latin American busways. Natural Resources Forum, JNRF 25 (2). 2005. Bus Rapid Transit, Sustainable transport: A sourcebook for policy-makers in developing cities. Eschborn: Deutsche Gesellschaft fur Technische Zusammenarbeit (GTZ) GmbH.

Wright, L. (2003). *Bus Rapid Transit.* Institute for Transportation and Development Policy. New York, USA

W. F. A. El-Washed (2001), A multi-objective transportation problem under fuzziness, *Fuzzy Sets and Systems*, Vol. 117, pp. 27–33.

Y. P. Aneja and K. P. K. Nair (1979), Bicriteria transportation problems, *Management Science*, Vol. 25, pp. 73–78.

4

Modelling, Simulation Methods for Intelligent Transportation Systems

George Papageorgiou and Athanasios Maimaris
European University Cyprus and University of Cyprus
Cyprus

1. Introduction

Effective transportation systems lead to the efficient movement of goods and people, which significantly contribute to the quality of life in every society. In the heart of every economic and social development, there is always a transportation system. Meanwhile, traffic congestion has been increasing worldwide because of increased motorization, urbanization, population growth, and changes in population density. This threatens the social and economic prosperity of communities all over the world. Congestion reduces utilization of the transportation infrastructure and increases travel time, air pollution, and fuel consumption. Therefore, managing and controlling transportation systems becomes a high priority task for every community, as it constitutes a matter of survival and prosperity for humanity.

In the search for meeting the demand for more traffic capacity, it has been realised repeatedly that building more roads is no longer a feasible solution due to the high cost and/or scarcity of land especially in metropolitan areas. In addition, the length of time that it takes to build additional roads and the disruption that this introduces to the rest of the traffic network makes the option of building new roads as the worst case scenario. The current highway transportation system runs almost open loop whereas traffic lights at surface streets are still lacking the intelligence that is necessary to reduce delays and speed up traffic flows. The recent advances in electronics, communications, controls, computers, and sensors provide an opportunity to develop appropriate transportation management policies and strategies in order to effectively utilize the existing infrastructure rather than building new road systems. The use of technologies will help provide accurate traffic data, implement control actions, and in general reduce the level of uncertainty and randomness that exists in today's transportation networks. The successful implementation of intelligent transportation systems will require a good understanding of the dynamics of traffic on a local as well as global system level and the effect of associated phenomena and disturbances such as shock wave generation and propagation, congestion initiation and so on. In addition, the understanding of human interaction within the transportation system is also crucial.

Transportation systems and traffic phenomena constitute highly complex dynamical problems where simplified mathematical models are not adequate for their analysis. There is a need for more advanced methods and models in order to analyse the causality, coupling, feedback loops, and chaotic behaviour involved in transportation problem situations. Traffic modelling can facilitate the effective design and control of today's complex transportation systems. Mathematical models cannot always accurately capture the high complexity and dynamicity of traffic systems. For this reason computer simulation models are developed and tuned to describe the traffic flow characteristics on a given traffic network. Once a computer simulation model is developed and validated using real data, different scenarios and new control strategies can be developed and simulated and evaluated before proposed for an actual implementation.

This chapter presents an overview of traffic flow modelling at the microscopic and macroscopic levels, a review of current traffic simulation software, as well as several methods for managing and controlling the various transportation system modes. In particular, section 2, examines the field of traffic flow theory and the concept of macroscopic vs. microscopic ways of modelling transportation systems. The derivation of traffic flow theory based on the law of conservation of mass, and the relationships between flow speed and density are presented in section 3 under the topic of macroscopic models. Section 4 analyses microscopic car following models and discusses advantages and limitations. Section 5 reviews various some of the most sophisticated traffic software modelling tools, all in relation to intelligent transportation systems. Finally, a summary of recent intelligent transportation systems studies carried out by the authors is provided in section 6 and conclusions are drawn in section 7.

2. Traffic flow modelling

The study of traffic flow (May, 1990), and in particular vehicular traffic flow, is carried out with the aim of understanding and assisting in the prevention and remedy of traffic congestion problems. The first attempts to develop a mathematical theory for traffic flow date back to the 1930s (Adams, 1937; Greenshields, 1935a), but despite the continuous research activity in the area we do not have yet a satisfactory mathematical theory to describe real traffic flow conditions. This is because traffic phenomena are complex and nonlinear, depending on the interactions of a large number of vehicles. Moreover, vehicles do not interact simply by following the laws of physics, but are also influenced by the psychological reactions of human drivers. As a result we observe chaotic phenomena such as cluster formation and backward propagating shockwaves of vehicle speed/density (Bose & Ioannou, 2000) that are difficult if at all possible to be accurately described with mathematical models. According to a state of the art report of the Transportation Research Board (Gartner, Messer, & Rathi, 2001), mathematical models for traffic flow may be classified as: Traffic Stream Characteristics Models, Human Factor Models, Car Following Models, Continuum Flow Models, Macroscopic Flow Models, Traffic Impact Models, Unsignalized Intersection Models, Signalized Intersection Models and Traffic Simulation Models. Below we describe briefly each of the above categories.

Traffic stream characteristics (Hall, 1996) theory involves various mathematical models, which have been developed to characterize the relationships among the traffic stream variables of speed, flow, and concentration or density.

Human factor modeling (Koppa, 1999), deals with salient performance aspects of the human element in the context of the human-machine interactive system. These include perception-reaction time, control movement time, responses to: traffic control devices, movement of other vehicles, hazards in the roadway, and how different segments of the population differ in performance. Further, human factors theory deals with the kind of control performance that underlies steering, braking, and speed control. Human factors theory provides the basis for the development of car following models. Car following models (Rothery, 1992), examine the manner in which individual vehicles (and their drivers) follow one another. In general, they are developed from a stimulus-response relationship, where the response of successive drivers in the traffic stream is to accelerate or decelerate in proportion to the magnitude of the stimulus. Car following models recognize that traffic is made up of discrete particles or driver-vehicle units and it is the interactions between these units that determine driver behavior, which affects speed-flow-density patterns. On the other hand, continuum models (Kuhne & Michalopoulos, 1997) are concerned more with the overall statistical behavior of the traffic stream rather than with the interactions between the particles. Following the continuum model paradigm, macroscopic flow models (J. C. Williams, 1997), discard the microscopic view of traffic in terms of individual vehicles or individual system components (such as links or intersections) and adopt instead a macroscopic view of traffic in a network. Macroscopic flow models consider variables such as flow rate, speed of flow, density and ignore individual responses of vehicles. Traffic impact models (Ardekani, Hauer, & Jamei, 1992) deal with traffic safety, fuel consumption and air quality models. Traffic safety models describe the relationship between traffic flow and accident frequency. Unsignalized intersection theory (Troutbeck & Brilon, 1997) deals with gap acceptance theory and the headway distributions used in gap acceptance calculations. Traffic flow at signalized intersections (Rouphail, Tarko, & Li, 1996) deals with the statistical theory of traffic flow, in order to provide estimates of delays and queues at isolated intersections, including the effect of upstream traffic signals. Traffic simulation modeling (Lieberman & Rathi, 1996) deals with the traffic models that are embedded in simulation packages and the procedures that are being used for conducting simulation experiments.

Mathematically the problem of modelling vehicle traffic flow can be solved at two main observation scales: the microscopic and the macroscopic levels. In the microscopic level, every vehicle is considered individually, and therefore for every vehicle we have an equation that is usually an ordinary differential equation (ODE). At a macroscopic level, we use the analogy of fluid dynamics models, where we have a system of partial differential equations, which involves variables such density, speed, and flow rate of traffic stream with respect to time and space.

The microscopic model involves separate units with characteristics such as speed, acceleration, and individual driver-vehicle interaction. Microscopic models may be classified in different types based on the so-called car-following model approach, as it will be discussed in section 4. The car-following modelling approach implies that the driver adjusts his or her acceleration according to the conditions of leading vehicles. In these models, the vehicle position is treated as a continuous function and each vehicle is governed by an ODE that depends on speed and distance of the car in the front. Another type of microscopic model involve the use of Cellular Automata or vehicle hopping models which differ from the car-following approach in that they are fully discrete time models. They consider the road as a string of cells that are either empty or occupied by one vehicle. One such model is the Stochastic Traffic Cellular Automata (Nagel, 1996; Nagel & Schreckenberg, 1992) model. Further, a more recent approach is currently under heavy research with the use of agent based modeling (Naiem, Reda, El-Beltagy, & El-Khodary, 2010).

Microscopic approaches are generally computationally intense, as each car has an ODE to be solved at each time step, and as the number of cars increases, so does the size of the system to be solved. Analytical mathematical microscopic models are difficult to evaluate but a remedy for this is the use of microscopic computer simulation. In such microscopic traffic models, vehicles are treated as discrete driver-vehicle units moving in a computer-simulated environment.

On the other hand, macroscopic models aim at studying traffic flow using a continuum approach, where it is assumed that the movement of individual vehicles exhibit many of the attributes of fluid motion. As a result, vehicle dynamics are treated as fluid dynamics. This idea provides an advantage since detailed interactions are overlooked, and the model's characteristics are shifted toward the more important parameters such as flow rate, concentration, or traffic density, and average speed, all being functions of one-dimensional space and time. This class of models is represented by partial differential equations. Modeling vehicular traffic via macroscopic models is achieved using fluid flow theory in a continuum responding to local or non-local influences. The mathematical details of such models are less than those of the microscopic ones. The drawback of macroscopic modeling is the assumption that traffic flow behaves like fluid flow, which is a rather harsh approximation of reality. Vehicles tend to interact among themselves and are sensitive to local traffic disturbances, phenomena that are not captured by macroscopic models. On the other hand, macroscopic models are suitable for studying large-scale problems and are computationally less intense especially after approximating the partial differential equation with a discrete time finite order equation.

There exists also a third level of analysis the so called mesoscopic level, which is somewhere between the microscopic and the macroscopic levels. In a mesoscopic or kinetic scale, which is an intermediate level, we define a function $f(t,x,v)$, which expresses the probability of having a vehicle at time t in position x at velocity v. This function, following methods of statistical mechanics, can be computed by solving an integro-differential equation, like the Boltzmann Equation (K. Waldeer, 2006; K. T. Waldeer, 2004).

The choice of the appropriate model depends on the level of detail required and the computing power available. Because of advancements in computer technology in recent years, the trend today is towards utilizing microscopic scale mathematical models, which incorporate human factors and car following models as a driver-vehicle behavior unit.

In the next two sections, macroscopic and microscopic models are examined in more detail.

2.1 Macroscopic traffic flow models

Macroscopic flow models (J. Williams, 1996), discard the real view of traffic in terms of individual vehicles or individual system components such as links or intersections and adopt instead a macroscopic fluid view of traffic in a network. In this section, we will cover the vehicle traffic flow fundamentals for the macroscopic modeling approach. The relationship between density, velocity, and flow is also presented. Then we derive the equation of conservation of vehicles, which is the main governing equation for scalar macroscopic traffic flow models. The macroscopic models for traffic flow, whether they are one-equation or a system of equations, are based on the physical principle of conservation. When physical quantities remain the same during some process, these quantities are said to be conserved. Putting this principle into a mathematical representation, it becomes possible to predict the density and velocity patterns at a future time.

Drawing an analogy between vehicle dynamics and fluid dynamics (Kuhne & Michalopoulos, 1997) let us consider a unidirectional continuous road section with two counting stations 1,2 at positions x_1 and x_2. The spacing between stations is Δx. In such a case, the number of cars in a segment of a highway Δx is a physical quantity, and the process is to keep it fixed so that the number of cars coming in equals the number of cars going out of the segment.

As it will be shown there is a close interrelationship between three traffic variables that is density, velocity and traffic flow. Suppose that in the above scenario, cars are moving with constant velocity v, and constant density ρ such that the distance d between the cars is also constant. Let an observer measure the number of cars N per unit time t that pass him/her (i.e. the traffic flow q).

Let N_1 be the number of cars passing station 1 and N_2 be the number of cars passing station 2 and Δt the duration of the observer counting time. Let q be the flow rate i.e. the number of cars passing a particular station per unit time, then

$$q_1 = \frac{N_1}{\Delta t} \text{ and } q_2 = \frac{N_2}{\Delta t}$$

$$\Delta q = \frac{N_2}{\Delta t} - \frac{N_1}{\Delta t} \text{ or } \Delta N = N_2 - N_1 = \Delta q \Delta t$$

For a build-up of cars therefore ΔN will be negative. Thus

$$\Delta N = (-\Delta q)\Delta t$$

Assuming that Δx is short enough so that vehicle density is uniform, then the increase in density during time Δt is given by

$$\Delta \rho = \frac{-\Delta N}{\Delta x} \text{ or } -\Delta N = \Delta \rho \Delta x$$

Assuming conservation of vehicles and no sinks or sources exist in the section of the roadway,

$$-\Delta q \Delta t = \Delta \rho \Delta x \text{ or } \frac{\Delta q}{\Delta x} + \frac{\Delta \rho}{\Delta t} = 0$$

Finally, assuming continuity of the medium and infinitesimal increments we get the conservation or continuity equation

$$\frac{\partial q}{\partial x}+\frac{\partial \rho}{\partial t}=0$$

In order to solve the above equation we assume that $v=f(\rho)$ and $q=\rho v$

$$\frac{\partial}{\partial x}(\rho v)+\frac{\partial \rho}{\partial t}=0 \text{ or } \frac{\partial}{\partial x}(\rho f(\rho))+\frac{\partial \rho}{\partial t}=0$$

Differentiating with respect to x ,

$$\frac{\partial \rho}{\partial x}f(\rho)+\frac{\partial f(\rho)}{\partial x}\rho+\frac{\partial \rho}{\partial t}=0 \text{ or } \frac{\partial \rho}{\partial x}f(\rho)+\frac{df\partial \rho}{d\rho \partial x}\rho+\frac{\partial \rho}{\partial t}=0$$

$$\Rightarrow (f(\rho)+\rho\frac{df}{d\rho})\frac{\partial \rho}{\partial x}+\frac{\partial \rho}{\partial t}=0$$

The above constitutes a first-order partial differential equation, which can be solved by the method of characteristics. A complete formulation and solution of the above equation has been published (Lighthill & Whitham, 1955). If the initial density and the velocity field are known, the above equation can be used to predict future traffic density. This leads us to choose the velocity function for the traffic flow model to be dependent on density and call it $V(\rho)$. The above equation assumes no generation or dissipation of vehicles. Sources and sinks may be added by including a function $g(x,t)$ on the RHS of the equation.

Several velocity-density-flow models have been developed through the years and are classified as single-regime or multi-regime models. Single-regime models assume a continuous relationship between velocity, density, and traffic flow while in multi-regime models the relationship is discontinuous depending on the density levels. Some of the most well-known single-regime and multi-regime models include the Greenshields model, Greenberg model, the Underwood model and Eddie's model. These are described as follows.

The Greenshields Model (Greenshields, 1935b) is a simple and widely used model. It is assumed that the velocity is a linearly decreasing function of the traffic flow density, and it is given by

$$v=v_f(1-\frac{\rho}{\rho_{jam}})$$

where v_f is the free flow speed and ρ_{jam} is the jam density. The above equation represents a monotonically decreasing function with respect to density. For zero density the model allows free flow speed v_f , while for maximum density ρ_{jam} we have 100% congestion where the speed is zero and no car is moving. Real traffic data shows that the speed-density relationship is indeed a rather linear negative slope function.

Based on the above empirical relationship Greenshields derived the following parabolic equation for the flow-speed-density relationship,

$$q = -(\frac{\rho_{jam}}{v_f})v^2 + \rho v$$

Real traffic data reflects somewhat the flow-density relationship for Greenshild's model, which follows a parabolic shape and shows that the flow increases to a maximum which occurs at some average density ρ and then it goes back to zero at high values of density. A third relationship that can be drawn is the speed flow function, which is again of parabolic shape.

Following Greenshields steps, Greenberg (Greenberg, 1959) developed a model of speed-density showing a logarithmic relationship. In Greenberg's model the speed-density function is given by

$$v = v_f - \ln(\frac{\rho}{\rho_{jam}})$$

Another single regime model is the Underwood model (Underwood, 1961), where the velocity-density function is represented as follows.

$$v = v_f e^{-\frac{\rho}{\rho_{jam}}}$$

(Edie, 1961) proposed a multi-regime model which basically combines Greenberg's and Underwood's model. Eddie suggested that

for densities $\rho \leq 50$

$$v = 54.9e^{-\frac{\rho}{163.9}}$$

and for densities $\rho \geq 50$

$$v = 28.6\ln(\frac{162.5}{\rho})$$

(Drake, Schofer, & May Jr, 1967) investigated seven speed-density models, including the above, through an empirical test in 1967. According to Drake et al the Eddie formulation gave the best estimates of the fundamental parameters but its root mean square error (rms) was the second lowest. The general conclusion though was that none of the models investigated provided a particularly good fit or explanation of the traffic data tested.

More recently, prominent researchers such as (Payne, 1979), (M. Papageorgiou, Blosseville, & Hadj-Salem, 1989) and (Michalopoulos, Yi, & Lyrintzis, 1992) developed macroscopic traffic flow simulation models based on a space-time discretization of the conservation equation. Even though these models are capable of describing complicated traffic phenomena with considerable accuracy, their main limitations arise in their inability to accurately simulate severe traffic congestion situations, where the conservation equation does not represent the traffic flow so well.

On the other hand, it is desirable to use macroscopic models if a good model can be found that satisfactorily describes the traffic flow for the particular traffic problem situation. The advantage of macroscopic models is their flexibility since detailed interactions are overlooked, and the model's characteristics are shifted toward important parameters such as flow rate, concentration or traffic density, and average speed. If the transportation/traffic problem demands more detail and accuracy such as the case of evaluating the effects of closely spaced intersections or bus priority systems on the traffic network then one should resort to microscopic models, which are described in the next section.

2.2 Microscopic driver-vehicle behaviour models

In this section well known car-following microscopic traffic flow models are presented and evaluated on their capability to realistically model traffic flow at the vehicle level.

In order to derive a one-dimensional simple car following model we first assume that cars do not pass each other. Then the idea is that a car in one-dimension can move and accelerate forward based on two parameters; the headway distance between the current car and the leading car, and their speed difference. Hence, it is called following, since a car from behind follows the one in the front.

Car-following models are based on the assumption that a stimulus response relationship exists that describes the control process of a driver-vehicle unit. This concept is expressed with the stimulus response equation where response is proportionally analogous to a stimulus based on a certain proportionality factor λ (Rothery, 1992) .

As seen later on in this section the various car following models incorporate a variant of the following stimulus response equation.

$$\ddot{x}_f(t) = \lambda \cdot \frac{\left(\dot{x}_l(t) - \dot{x}_f(t)\right)}{\left(x_l(t) - x_f(t)\right)}$$

where x_f is the one-dimensional position of the following vehicle, x_l is the one-dimensional position of the leading vehicle, and t is time. The response function here is taken as the acceleration of the following vehicle as the driver experiences inertia forces and has direct control on acceleration/deceleration through the accelerator and brake pedals.

The above stimulus-response equation of car-following is a simplified description of a complex phenomenon. A generalization of car following in a conventional control theory block diagram is shown in figure 1. As seen in figure 1 a more complete representation of car following would include a set of equations that are able to model the dynamical properties of the vehicle and the roadway characteristics. It would also include the psychological and physiological properties of drivers, as well as couplings between vehicles, other than the forward nearest neighbours and other driving tasks such as lateral control, the state of traffic, and emergency conditions and other factors.

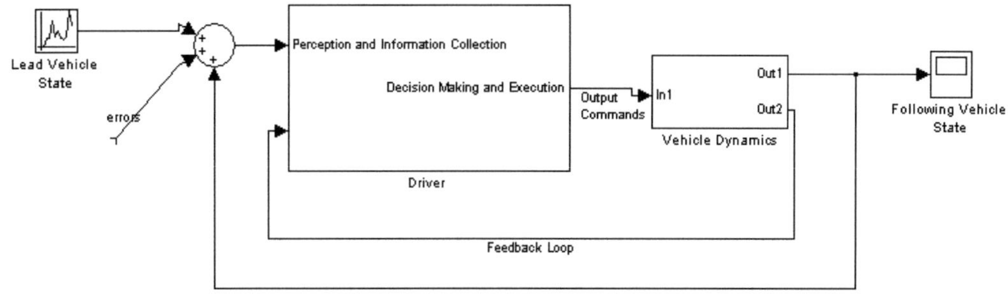

Fig. 1. A Generalised Block Diagram of Car Following (adapted from Rothery, 1992).

The car-following behaviour is basically, a human interactive process where the driver of the vehicle attempts to reach a stable situation and maintain it by following a leading vehicle, by continuously taking corrective actions like accelerating or decelerating. As it will be seen in the next paragraphs car following models may be classified as Stimulus-Response models, safety distance or collision avoidance models, psychophysical or action point models, and fuzzy logic models. Some of the most widely applied car following models are presented below.

Pipes (Pipes, 1967) proposed a theory of car following behaviour based on what he referred to as the "idealized law of separation". The law specifies that each vehicle must maintain a certain prescribed "following distance" from the preceding vehicle. This distance is the sum of a distance proportional to the velocity of the following vehicle and a certain given minimum distance of separation when the vehicles are at rest. Such a model implies that the actions of the following vehicle are only affected by the relative speed between the leading vehicle and the following vehicle. Forbes (Forbes, 1963) modelled car following behaviour by assuming that drivers choose to keep a minimum time gap from the rear end of leading vehicle. The Forbes model of car following also implies that the actions of the following vehicle are only affected by the relative speed between the leading vehicle and the following vehicle.

The General Motors Research Laboratories published significant amount of work on the car-following theory model in a series of papers (Gazis, Herman, & Potts, 1959; Herman, Montroll, Potts, & Rothery, 1959). The basic idea used here is that the actions of the following vehicle in terms of acceleration or deceleration are a function of a single stimulus and the sensitivity of the following vehicle to the stimulus under the prevailing conditions. The stimulus is assumed to be the relative speed between leading and the following vehicle. Sensitivity to the stimulus is assumed to be affected by the distance headway between the leading vehicle and the following vehicle as well as the speed of following vehicle.

Other approaches such as those by Rockwell et al (Rockwell, Ernest, & Hanken, 1968) present a regression based car-following model, which takes into consideration two leading vehicles, and Chakroborty and Kikuchi (Kikuchi, Chakroborty, & Engineering, 1992) a Fuzzy Inference based car-following model.

Consolidating the various approaches of car following models it can be concluded that the general assumption about the interaction between a leader and follower car is governed by the following equation (Rothery, 1992).

$$a_f = F(v_l, v_f, s, d_l, d_f R_f P_i)$$

where a_f is the acceleration (response) of the following vehicle, v_l is the velocity of the leading vehicle, v_f is the velocity of the following vehicle, s is the spacing between follower and leader vehicles or separation distance, d_l and d_f are the projected deceleration rates of the leader and follower vehicles, respectively, R_f is the reaction time of the driver in the following vehicle and P_i are other parameters specific to the car-following model. Based on this generalised equation several car following models have been developed through the years and are briefly presented below using the above general model notation.

(Chandler, Herman, & Montroll, 1958) developed a car following model assuming that the following vehicle driver responds solely to changes in the speed of the immediate leader vehicle. Chandler's model is given by the following equation.

$$a_f = P_1(v_l - v_f)$$

(Gazis, Herman, & Rothery, 1961) developed a more complex model adding the position of the leading and following vehicles in the equation and thereby introducing the notion of a safety distance between the two vehicles.

$$a_f = \frac{P_1}{d_l - d_f}(v_l - v_f)$$

(Edie, 1961) developed a similar response-stimulus model involving velocity and position changes as shown below.

$$a_f = P_1 \frac{V_f}{(d_l - d_f)^2}(V_l - V_f)$$

(Herman & Rothery, 1962) apart from velocity and position changes included some other parameters in their car following model.

$$a_f = P_1 \frac{V_f^{P_2}}{(d_l - d_f)^{P_3}}(V_l - V_f)$$

(Bierley, 1963) in a similar fashion but increasing the number of parameters suggested the following model.

$$a_f = P_1(v_l - v_f) + P_2(d_l - d_f)$$

(Fox & Lehman, 1967) added another lead vehicle in their car following model by considering the changes in speed and position of the vehicle in front of the immediate leading vehicle as shown below.

$$a_f = P_1 V_f \left(\frac{P_2(v_{ll} - v_f)}{(d_{ll} - d_f)^2} + \frac{P_3(v_l - v_f)}{(d_l - d_f)^2}\right)$$

where v_{ll} the speed of the vehicle in front of the immediate leading vehicle and d_{ll} the separation distance of vehicle in front of the immediate leading vehicle.

Bexelius (Bexelius, 1968) also suggested a car following model that takes into account two leading vehicles as follows.

$$a_f = P_1(v_{ll} - v_f) + P_2(v_l - v_f)$$

More complex car following models of similar nature were developed by (Wicks, Lieberman, Associates, & Division, 1980) where the NETSIM software is based, and by (Bullen, 1982) where FRESIM is based. Further, (Gipps, 1981) developed a safety distance or behavioural model, which is employed by AIMSUN. (Fritzsche, 1994) and (Wiedemann, 1974) developed the so called action point or phycho-physical models for Paramics and VISSIM respectively. Some more modern approaches to car following models make use of fuzzy logic algorithms (Gonzalez-Rojo, Slama, Lopes, & Mora-Camino, 2002; Yikai, Satoh, Itakura, Honda, & Satoh, 1993). Interesting is also the employment of the System Dynamics modelling principles in car following. More recently, (Mehmood, Saccomanno, & Hellinga, 2003) introduced the system dynamics method into a successful car following model, which takes into consideration the interactions of a following and two leading vehicles. Further, other techniques from artificial intelligence are being utilised in an effort to make car following models more realistic.

Even though there were many efforts through the years to develop realistic models of car following behaviour there are significant limitations concerning their validity. Limitations arise mainly from unrealistic assumptions about the ability of drivers of following vehicles to perceive relative or absolute speeds and accelerations of the interacting vehicles. As Boer (Boer, 1999) suggests factors such as aging impairment and disability further influence driver reactions, which current car following models do not take into consideration. Further, unrealistic is the assumption that driver behaviour is influenced only by the immediate leading vehicle motion as observed by a number of researchers such as Fox and Lehman, and Bexelius. Also the assumption for an empirical relationship fails to explain actual behaviour as pointed out by (Van Winsum, 1999) and (Gipps, 1981) . Finally, existing car following models are rather idealistic as they assume symmetrical driver responses to traffic stimuli, which is clearly unrealistic as revealed by (Chakroborty & Kikuchi, 1999).

As (Brackstone & McDonald, 1999) conclude in their review on microscopic car following models there are potential pitfalls awaiting the unwary in the use of microcopic models. A comprehensive review on the weaknesses and potential developments of microscopic models is given by (Brackstone & McDonald, 1999).

Research on the existing models of driver behaviour has been restricted to modelling driver behaviour under car-following situations. Little work was found, on models of driver behaviour under various other driving situations. It can be said that most research so far has been concentrated on modelling driver behaviour in situations, where only longitudinal interactions affect the driver. Situations where, either lateral interactions alone or lateral as well as longitudinal interactions affect driver's behaviour received much less attention.

3. Microscopic traffic modelling software tools

Microscopic simulation is a term used in traffic modelling and is typified by software packages such as VISSIM (Fellendorf & Vortisch, 2001; Gomes, May, & Horowitz, 2004; Park, Won, & Yun, 2006; PTV, 2005), CORSIM (Lin, 1998; Prevedouros & Wang, 1999; Zhang, McHale, & Zhang, 2003), and PARAMICS (Gardes, 2006; Jacob & Abdulhai, 2006; Ozbay, Bartin, Mudigonda, & Board, 2006). Traffic simulation microscopic models simulate the behaviour of individual vehicles within a predefined road network and are used to predict the likely impact of changes in traffic patterns resulting from proposed commercial developments or road schemes. They are aiming to facilitate transportation consultants, municipalities, government transportation authorities and public transportation companies. The traffic flow models used are discrete, stochastic, time step based microscopic models, with driver-vehicle units as single entities.

Traffic simulation software modelers combine in a single package multiple traffic flow mathematical models and therefore make it possible to combine the current knowledge on traffic theory when analyzing a traffic congestion problem. A screenshot of the VISSIM graphical user interface is provided in figure 2. The microscopic model depicted in the figure was developed in order to analyze traffic and evaluate the impact of various bus priority scenarios for a traffic network in Nicosia, Cyprus (G. Papageorgiou, 2006; G. Papageorgiou, Damianou, Pitsillides, Aphames, & Ioannou, 2006).

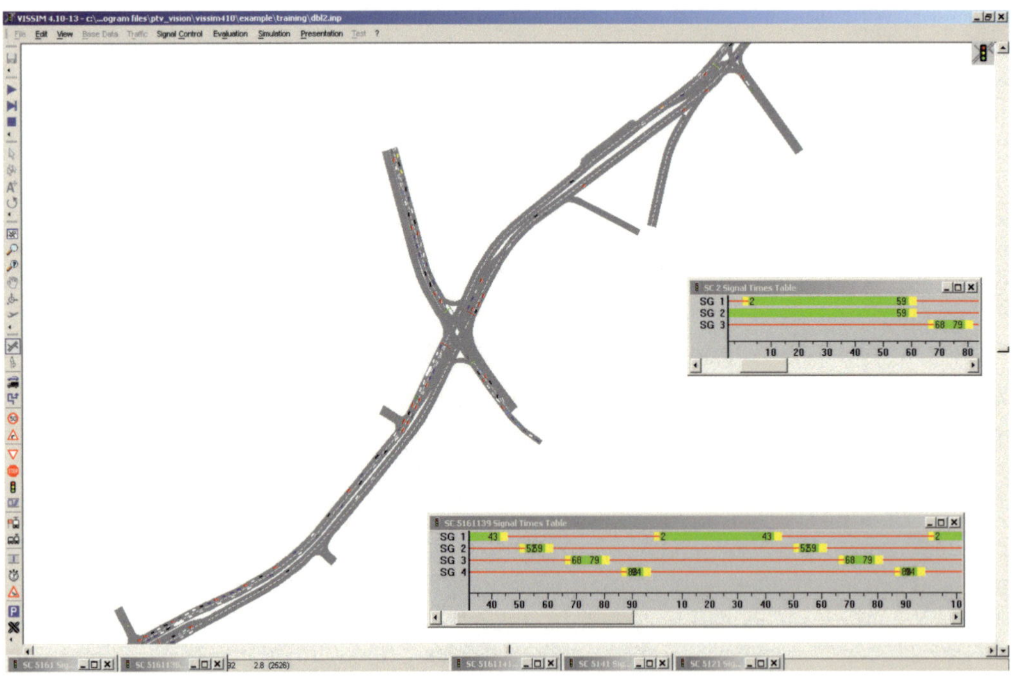

Fig. 2. VISSIM Graphical User Interface depicting part of the microscopic model of Strovolos Ave. in Nicosia, Cyprus.

In today's traffic simulation software, data such as network definition of roads and tracks, technical vehicle and behavioural driver specifications, car volumes and paths can be inserted in graphical user interface mode. Values for acceleration, maximum speed and desired speed distributions can be configured by the user to reflect local traffic conditions. Various vehicles types can also be defined. Further, traffic control strategies and algorithms may be defined as well as interfaces may be built with well-known urban traffic controllers. CORSIM, PARAMICS, VISSIM, and AIMSUN were calibrated and validated in a number of traffic studies worldwide. Below we present some of their main features.

CORSIM which stands for Corridor microscopic simulation is developed by Federal Highway Administration of United States. It has evolved from two separate traffic simulation programs NETSIM and FRESIM. NETSIM models arterials with signalised and unsignalised intersections, while FRESIM models uninterrupted freeways and urban highways.

In the case of VISSIM the microscopic model consists of a psycho-physical car following model for longitudinal vehicle movement and a rule-based lane changing algorithm for lateral movements. The model is based on an urban and a freeway model which were developed by Wiedemann from the University of Karlsruhe. VISSIM is especially well known for its signal control module, which uses a vehicle actuated programming language can model almost any traffic control logic. Further, VISSIM scores high on its ability to model public transportation systems.

AIMSUN was developed by TSS in order to simulate urban and interurban traffic networks. It is based on the car-following model of Gibbs . AIMSUN is therefore based on a collision avoidance car-following model. Traffic can be modelled via input flows and turning movements, origin destination matrices, and route choice models.

PARAMICS, which stands for Parallel Microscopic Simulation, comprises of various modules which include a modeller, a processor, an analyser, a monitor, a converter and an estimator. PARAMICS is well known for its visualization graphics and for its ability to model quite a diverse range of traffic scenarios.

A comprehensive review of simulation models of traffic flow was conducted by the Institute for Transport Studies at the University of Leeds as part of the SMARTEST Project which is a collaborative project to develop micro-simulation tools to help solve road traffic management problems. The study compared the capabilities of more than 50 simulation packages. The results are available on the internet at http://www.its. leeds.ac.uk/ projects/smartest. Other significant reviews of traffic simulation software include the work of (Bloomberg & Dale, 2000) who compared Corsim and Vissim as well as the work of (Boxill, Yu, Training, Research, & Center, 2000) who compared the capabilities of Corsim, Aimsun and Paramics. It can be concluded from the various reviews that software modelers that have comparative capabilities include VISSIM, AIMSUN, and PARAMICS.

In a more recent comparative study of microscopic car following behavior, (Panwai & Dia, 2005) evaluate AIMSUN, VISSIM and PARAMICS. They concluded that the accuracy of a traffic simulation system depends highly on the quality of its traffic flow model at its core, which consists of car following and lane changing models. In the study the car-following behaviour for each simulator was compared to field data obtained from instrumented vehicles travelling on an urban road in Germany. The Error Metric on distance (Manstetten, Krautter, & Schwab, 1997) performance indicator gave substantially better values for AIMSUN than those of VISSIM and PARAMICS. Further, the Root Mean Square Error (RMSE) was substantially less for VISSIM and AIMSUN than the RMSE for PARAMICS. In another paper presented at the 9th TRB Conference on the Application of Transportation Planning Methods Choa et al. (Choa, Milam, & Stanek, 2004) concluded that although CORSIM provides the shortest traffic network setup time , PARAMICS and VISSIM generated simulation results that better matched field observed conditions and traffic engineering principles.

The reason microscopic simulation models are used over other software packages and methods like the Highway Capacity Manual (HCM) is that microscopic simulations allow us to evaluate the effects that different traffic elements have on each other. Being able to evaluate the effects of closely spaced intersections and interchanges on the traffic network or the effects of a bottleneck condition on the surrounding system, can only be achieved by microscopic traffic simulation models. Also, as metropolitan traffic conditions experience congestion over 3 to 4 hour periods, microscopic traffic simulation programs allow us to evaluate the build-up to congested conditions and the recovery of the system at the end of the period. The peak period of congestion is complex and evaluating solutions under these conditions can only be accomplished using microscopic simulation tools.

In the following section, an approach to modelling and simulation of intelligent transportation systems (ITS) is proposed and implemented for a particular case study in Nicosia , Cyprus. The approach utilizes the VISSIM microscopic simulation modeler.

4. ITS studies using microscopic simulation

As described in the previous sections traffic phenomena constitute a dynamical problem situation, which makes traffic modeling and simulation a very complex, iterative and tedious process. In order to increase chances for developing a realistic simulation model the following methodology is developed, which is based on the suggestions of (Lieberman & Rathi, 1996) and (Dowling, 2007). This is applied for the modeling of Archangelou Avenue traffic network in Nicosia, Cyprus as described below.

The study area is depicted in figure 3, which shows Archangelou Avenue with its nearby traffic network. Archangelou Avenue is the main road connecting the Rural Nicosia District to the centre of Nicosia, where the main business center is located. Nicosia, the capital of Cyprus, has a population of around 350,000 people. Archangelou Ave., which is one of the three main arterial roads exhibits very high traffic flows as compared with the other regions of metropolitan Nicosia. Further, Archangelou Avenue serves as the connector between Nicosia and a large and heavily populated area of urban and rural communities.

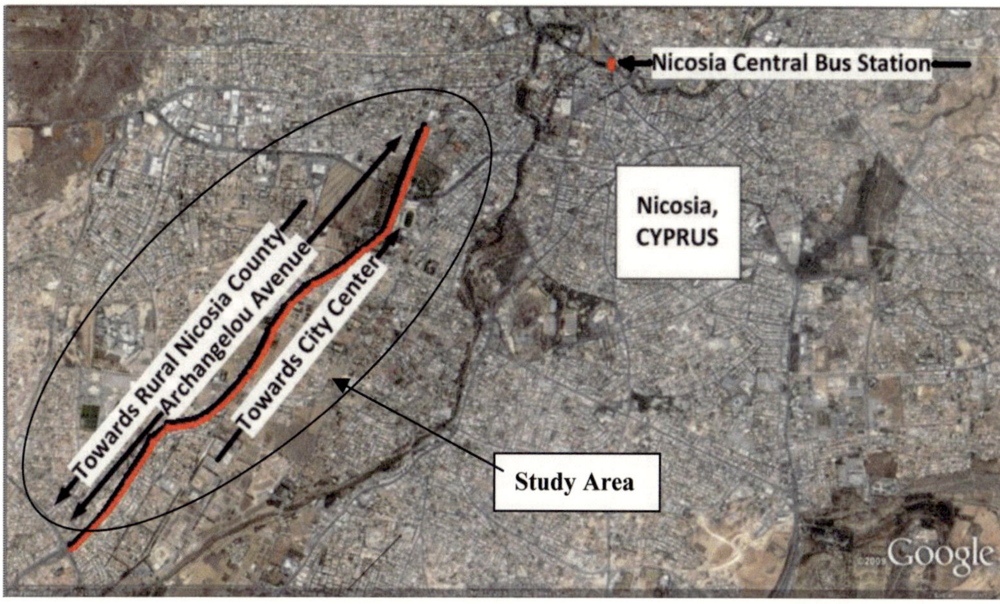

Fig. 3. The Archangelou Avenue Case Study Area with the nearby Traffic Network.

The aim of the study titled "Intelligent Transportation Systems in Archangelou Avenue (BUSSIM)" was to develop and test BRT strategies via scenario analysis in a computer simulated environment. Scenarios that were evaluated include a number of configurations regarding the introduction of dedicated bus lanes as well as bus advance signal areas as well as High Occupancy Vehicle (HOV) lanes. The scenario analysis was carried out via computer experiments using a microscopic simulation model of Archangelou Avenue urban traffic network. The case study presented in this chapter is part of the BUSSIM research project (George Papageorgiou, Maimaris, Ioannou, Pitsillides, & Afamis, 2010) which was funded by the Cyprus Research Promotion Foundation and Transim Transportation Research Ltd.

As shown in figure 4, the first step of the proposed approach is to identify and define the problem. In our case the symptoms of the problem which are attributed to traffic congestion manifest themselves as increasing travel times for all transport modes. The main causes to the problem of traffic congestion in Nicosia consist of an increasing number of vehicles and a decreasing use of the bus transportation system. Adding more capacity to the road infrastructure will only make things worse, as a reinforcing feedback loop is created where further use of private vehicles is encouraged and use of the public transport is discouraged. Therefore, the long term solution to the problem is to balance or even to turn around the situation by encouraging the use of the public transport mode.

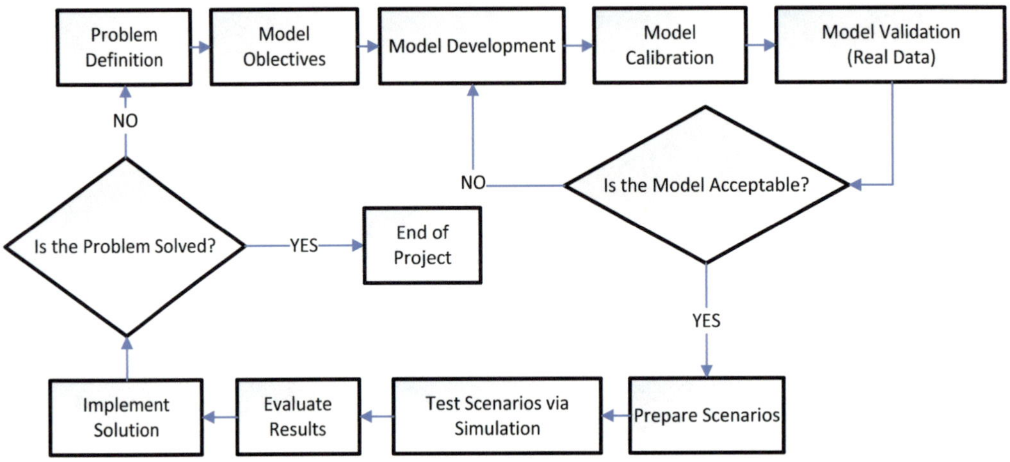

Fig. 4. The Proposed Traffic Modeling and Simulation Method (in paper by Papageorgiou et al presented at 12th IFAC Symposium on Transportation Systems, September 2009).

The question then becomes how to attract people in using the bus transportation system. The answer to this was given by the citizens of Cyprus in a recent survey where they expressed their wish for higher quality, faster public bus transport system. This is what was investigated in the BUSSIM project, concentrating on providing a faster and better quality level of service for bus passengers. The objective therefore in the modeling and simulation method was to examine various scenarios such as dedicated bus lanes and Bus Rapid Transit Systems that would provide a better level of service for the bus transportation system. Meanwhile there was a need to anticipate and assess any side effects of plausible solutions to the rest of the transportation system.

Based on the stated model objectives, a microscopic simulation model of Archangelou Avenue was developed. Like any other traffic network, Archangelou Avenue consisted of many traffic parameters that needed to be taken into account. These included traffic control signals, priority rules, routing decisions, and pedestrian crossings, signalized and un-signalized intersections and so on. A helicopter view of the simulation model of Archangelou Avenue is depicted in figure 5 (see also figure 3). Figure 5 shows the proposed layout of Archangelou Ave, which is a five-lane road more than 4 kilometers long, as well as the main roads that intersect Archangelou Avenue. Figure 5 also shows potential areas for introducing dedicated bus lanes.

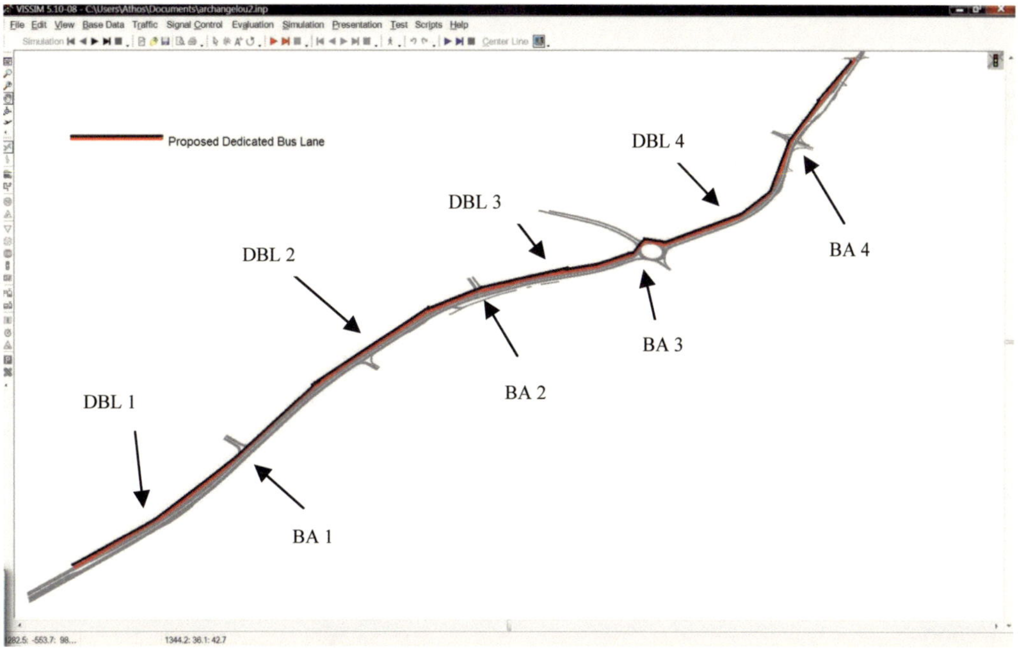

Fig. 5. The Traffic Simulation Model Showing The Proposed Road Design for Dedicated Bus Lane (DBL) and Bus Advance Areas (BA).

The model incorporated a significant amount of various traffic data that may be classified in terms of static data and dynamic data. Static data represents the roadway infrastructure. It includes links, which are directional roadway segments with a specified number of lanes, with start and end points as well as optional intermediate points. Further, static data includes connectors between links, which are used to model turnings, lane drops and lane gains, locations and length of transit stops, position of signal heads/stop lines including a reference to the associated signal group, and positions and length of detectors. Dynamic data was also specified for the traffic simulation experiments. It included traffic volumes with vehicle mix for all links entering the network, locations of route decision points with routes, that is the link sequences to be followed, differentiated by time and vehicle classification, priority rules, right-of-way to model un-signalized intersections, permissive turns at signalized junctions and yellow boxes or keep-clear-areas, locations of stop signs, public transport routing, departure times and dwell times.

Having introduced the necessary traffic parameters in the model, the iterative process begun, which consisted of model development calibration and validation of the model.

Figure 6 shows the real Vs simulated traffic flows of the various vehicle movement directions of a central intersection of Archangelou Avenue, in particular that of Archangelou-Odyssea Elyti. As seen in the bar chart, traffic flows of real measurements obtained and those of simulated results, are quite comparable. In particular the error ranges from only 1% to 5%, a fact that contributes to building confidence for the model. Further, the simulation model demonstrated the queues that are encountered in reality during the morning peak hours.

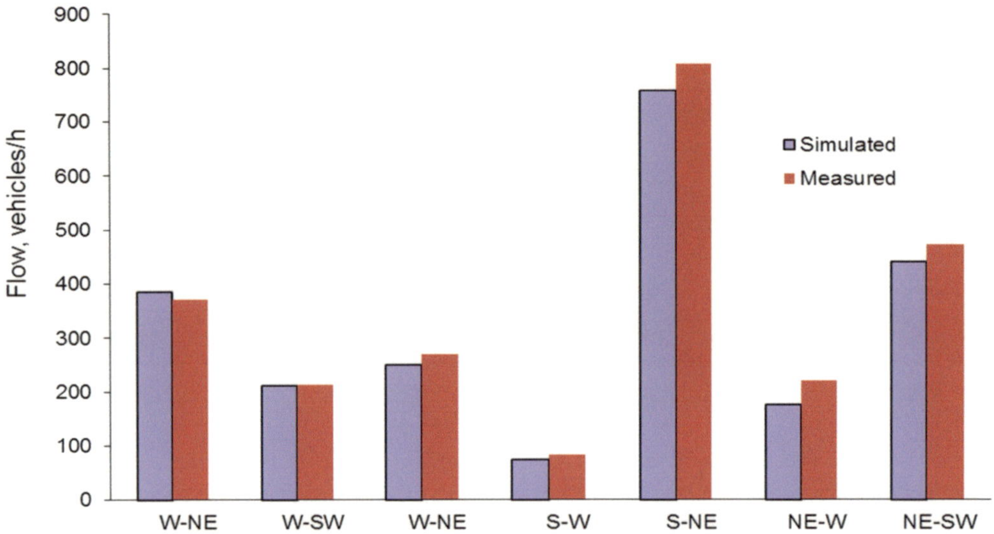

Fig. 6. Model Validation: Simulated Vs Measured flows at Archangelou-Odyssea Elyti junction.

With a validated model in our hands, next comes the preparation of BRT scenarios, their evaluation and the analysis of the results. After consultations with the transportation planning section of the Ministry of Communications Works we came up with several plausible scenarios. In summary, the various scenarios involve the use of dedicated bus lanes and High Occupancy Vehicle lanes by means of Intelligent Transportation Systems.

Even though the modelling process and especially the calibration of the microscopic model was time consuming, results from the simulation experiments gave us significant information on a variety of measures of effectiveness (MoE). In particular, we have managed to compute travel times, queue lengths, delays, average speeds, lane changes and other MoEs for the various scenarios under investigation. On the basis of the various MoEs comparison was carried out between the various scenarios using hypothesis testing with a 95% confidence interval (results submitted for publication). Such valuable information is obviously essential for the implementation of any Intelligent Transportation Systems project.

5. Conclusion

Modelling and simulation methods are essential elements in the design, operation and control of Intelligent Transportation Systems (ITS). Congestion problems in cities worldwide have drawn a high level of interest for better management and control of transportation systems. Of major importance are ITS systems that include advanced traffic management and control techniques. Such techniques include real-time traffic control measures and real-time traveller information and guidance systems whose purpose is to assist travellers in making departure time, mode and route choice decisions. Transportation research is heading towards developing models and simulators for use in the planning, design and operations and control of such intelligent transportation systems.

This chapter presented an overview of the most important developments in traffic flow theory, and examines modelling of traffic flow at two fundamental levels: the macroscopic level, where traffic is regarded as a fluid, and the microscopic level, where traffic is represented by individual driver-vehicle units. Concerning these two levels of analysis, without discarding the usefulness of macroscopic models it may be concluded that as a result of advancements in computer technology, and the need for more detailed and accurate traffic models there is a trend nowadays for microscopic traffic models where the ultimate goal is to capture the driver-vehicle unit interactions under a variety of driving conditions in a computer simulated environment.

Further, this chapter provided an insight analysis to the world's most sophisticated traffic simulation modeller software, VISSIM, AIMSUN, CORSIM and PARAMICS, where their capabilities and limitations are discussed. Also, an approach to modeling and simulation of intelligent transportation systems is proposed and implemented. The proposed approach goes through various stages, which include problem identification, model objectives, model development, model calibration, model validation, scenario preparation, simulation experiments and simulated results evaluation. The proposed approach is applied in the case of developing a microscopic traffic simulation model for the urban traffic network of Archangelou Avenue, Nicosia, Cyprus in order to examine alternative bus transport mode enhancements by means of Intelligent Transportation Systems.

6. References

Adams, W. F. (1937). *Road traffic considered as a random series*: Institution of Civil Engineers.

Ardekani, S., Hauer, E., & Jamei, B. (1992). Traffic impact models. *Chapter 7 in Traffic Flow Theory, Oak Bridge National Laboratory Report.*

Bexelius, S. (1968). An extended model for car-following. *Transportation Research, 2*(1), 13-21.

Bierley, R. L. (1963). Investigation of an Intervehicle Spacing Display. *Highway Research Record.*

Bloomberg, L., & Dale, J. (2000). Comparison of VISSIM and CORSIM traffic simulation models on a congested network. *Transportation Research Record: Journal of the Transportation Research Board, 1727*(-1), 52-60.

Boer, E. R. (1999). Car following from the driver's perspective. *Transportation Research Part F: Traffic Psychology and Behaviour, 2*(4), 201-206.

Bose, A., & Ioannou, P. (2000). *Shock Waves in Mixed Traffic Flow.*

Boxill, S. A., Yu, L., Training, T. S. U. C. f. T., Research, & Center, S. R. U. T. (2000). *An evaluation of traffic simulation models for supporting ITS development*: Center for Transportation Training and Research, Texas Southern University.

Brackstone, M., & McDonald, M. (1999). Car-following: a historical review. *Transportation Research Part F: Traffic Psychology and Behaviour, 2*(4), 181-196.

Bullen, A. (1982). Development of Compact Microsimulation for Analyzing Freeway Operations and Design. *Transportation Research Record*(841).

Chakroborty, P., & Kikuchi, S. (1999). Evaluation of the General Motors based car-following models and a proposed fuzzy inference model. *Transportation Research Part C: Emerging Technologies, 7*(4), 209-235.

Chandler, R. E., Herman, R., & Montroll, E. W. (1958). Traffic dynamics: studies in car following. *Operations Research,* 165-184.

Choa, F., Milam, R. T., & Stanek, D. (2004). *CORSIM, PARAMICS, and VISSIM: What the Manuals Never Told You.*

Dowling, R. (2007). Traffic Analysis Toolbox Volume VI: Definition, Interpretation, and Calculation of Traffic Analysis Tools Measures of Effectiveness: Federal Highway Administration Report FHWA-HOP-08-054.

Drake, J., Schofer, J., & May Jr, A. (1967). A statistical analysis of speed density hypotheses, Highway Res: Record.

Edie, L. C. (1961). Car-following and steady-state theory for noncongested traffic. *Operations Research*, 66-76.

Fellendorf, M., & Vortisch, P. (2001). *Validation of the microscopic traffic flow model VISSIM in different real-world situations.*

Forbes, T. (1963). Human factor considerations in traffic flow theory. *Highway Research Record.*

Fox, P., & Lehman, F. G. (1967). *Safety in car following: a computer simulation*: Newark College of Engineering.

Fritzsche, H. T. (1994). A model for traffic simulation. *Traffic Engineering and Control, 35*(5), 317-321.

Gardes, Y. (2006). *Evaluating Traffic Calming and Capacity Improvements on SR-20 Corridor Using Microscopic Simulation.*

Gartner, N., Messer, C. J., & Rathi, A. K. (2001). Traffic flow theory: A state-of-the-art report. *Transportation Research Board, Washington DC.*

Gazis, D. C., Herman, R., & Potts, R. B. (1959). Car-following theory of steady-state traffic flow. *Operations Research*, 499-505.

Gazis, D. C., Herman, R., & Rothery, R. W. (1961). Nonlinear follow-the-leader models of traffic flow. *Operations Research*, 545-567.

Gipps, P. G. (1981). A behavioural car-following model for computer simulation. *Transportation Research Part B: Methodological, 15*(2), 105-111.

Gomes, G., May, A., & Horowitz, R. (2004). Congested freeway microsimulation model using VISSIM. *Transportation Research Record: Journal of the Transportation Research Board, 1876*(-1), 71-81.

Gonzalez-Rojo, S., Slama, J., Lopes, P. A., & Mora-Camino, F. (2002). A fuzzy logic approach for car-following modelling. *Systems Analysis Modelling Simulation, 42*(5), 735-755.

Greenberg, H. (1959). An analysis of traffic flow. *Operations Research*, 79-85.

Greenshields, B. (1935a). *A study in highway capacity, highway research board.*

Greenshields, B. (1935b). A study in highway capacity. Highway Res. *Board Proc. v14*, 448-477.

Hall, F. L. (1996). Traffic stream characteristics. *Traffic Flow Theory. US Federal Highway Administration.*

Herman, R., Montroll, E. W., Potts, R. B., & Rothery, R. W. (1959). Traffic dynamics: analysis of stability in car following. *Operations Research*, 86-106.

Herman, R., & Rothery, R. (1962). Microscopic and macroscopic aspects of single lane traffic flow. *Operations Research, Japan*, 74.

Jacob, C., & Abdulhai, B. (2006). Automated adaptive traffic corridor control using reinforcement learning: approach and case studies. *Transportation Research Record: Journal of the Transportation Research Board, 1959*(-1), 1-8.

Kikuchi, S., Chakroborty, P., & Engineering, U. o. D. D. o. C. (1992). *A car-following model based on fuzzy inference system*: Transportation Research Board, National Research Council.

Koppa, R. J. (1999). Human factors. *Traffic Flow Theory, a state of the art report. Revised Monograph on Traffic Flow Theory", ed. by: Gartner, N., CJ Messer & AK Rathi.*

Kuhne, R., & Michalopoulos, P. (1997). Continuum flow models. *Traffic flow theory: A state of the art reportrevised monograph on traffic flow theory.*

Lieberman, E., & Rathi, A. (1996). Traffic Simulation. in Traffic Flow Theory. *Washington, DC: US Federal Highway Administration*, 10-11.

Lighthill, M. J., & Whitham, G. B. (1955). On kinematic waves. II. A theory of traffic flow on long crowded roads. *Proceedings of the Royal Society of London. Series A. Mathematical and Physical Sciences, 229*(1178), 317.

Lin, S. (1998). CORSIM micro-node logic: Technical Report, Federal Highway Administration, McLean, VA.

Manstetten, D., Krautter, W., & Schwab, T. (1997). *Traffic simulation supporting urban control system development.*

May, A. D. (1990). *Traffic flow fundamentals*: Prentice Hall.

Mehmood, A., Saccomanno, F., & Hellinga, B. (2003). Application of system dynamics in car-following models. *Journal of transportation engineering, 129*, 625.

Michalopoulos, P. G., Yi, P., & Lyrintzis, A. S. (1992). Development of an improved high-order continuum traffic flow model. *Transportation Research Record*(1365).

Nagel, K. (1996). Particle hopping models and traffic flow theory. *Physical Review E, 53*(5), 4655.

Nagel, K., & Schreckenberg, M. (1992). A cellular automaton model for freeway traffic. *Journal de Physique I, 2*(12), 2221-2229.

Naiem, A., Reda, M., El-Beltagy, M., & El-Khodary, I. (2010) *An agent based approach for modeling traffic flow.*

Ozbay, K., Bartin, B. O., Mudigonda, S., & Board, T. R. (2006). *Microscopic Simulation and Calibration of Integrated Freeway and Toll Plaza Model.*

Panwai, S., & Dia, H. (2005). Comparative evaluation of microscopic car-following behavior. *Intelligent Transportation Systems, IEEE Transactions on, 6*(3), 314-325.

Papageorgiou, G. (2006). *Towards a microscopic simulation model for traffic management: a computer based approach.*

Papageorgiou, G., Damianou, P., Pitsillides, A., Aphames, T., & Ioannou, P. (2006). *A Microscopic Traffic Simulation Model for Transportation Planning in Cyprus.*

Papageorgiou, G., Maimaris, A., Ioannou, P., Pitsillides, A., & Afamis, T. (2010). *Introduction of Bus Rapid Transit in Cyprus: Evaluation of Bus Priority Scenarios.* Paper presented at the Transportation Research Board 89th Annual Meeting.

Papageorgiou, M., Blosseville, J. M., & Hadj-Salem, H. (1989). Macroscopic modelling of traffic flow on the Boulevard Périphérique in Paris. *Transportation Research Part B: Methodological, 23*(1), 29-47.

Park, B., Won, J., & Yun, I. (2006). Application of microscopic simulation model Calibration and validation procedure: Case study of coordinated actuated signal system. *Transportation Research Record: Journal of the Transportation Research Board, 1978*(-1), 113-122.

Payne, H. J. (1979). FREFLO: A macroscopic simulation model of freeway traffic. *Transportation Research Record*(722).

Pipes, L. A. (1967). Car following models and the fundamental diagram of road traffic. *Transportation Research, 1*(1), 21-29.

Prevedouros, P. D., & Wang, Y. (1999). Simulation of large freeway and arterial network with CORSIM, INTEGRATION, and WATSim. *Transportation Research Record: Journal of the Transportation Research Board, 1678*(-1), 197-207.

PTV (2005). VISSIM Version 4.10. *User Manual, March.*

Rockwell, T. H., Ernest, R. L., & Hanken, A. (1968). A Sensitivity Analysis of Empirical Car-Following Models. *Transportation Research, 2*, 363-373.

Rothery, R. W. (1992). Car following models. *Trac Flow Theory.*

Rouphail, N., Tarko, A., & Li, J. (1996). Traffic Flow at Signalized Intersections. in Traffic Flow Theory. *Washington, DC: US Federal Highway Administration*, 9-1.

Troutbeck, R., & Brilon, W. (1997). Unsignalized Intersection Theory, Revised Traffic Flow Theory.

Underwood, R. (1961). Speed, volume, and density relationships: Quality and theory of traffic flow. *Yale bureau of highway traffic*, 141–188.

Van Winsum, W. (1999). The human element in car following models. *Transportation Research Part F: Traffic Psychology and Behaviour, 2*(4), 207-211.

Waldeer, K. (2006). *Kinetic Theory in Vehicular Traffic Flow Modeling.*

Waldeer, K. T. (2004). Numerical investigation of a mesoscopic vehicular traffic flow model based on a stochastic acceleration process. *Arxiv preprint cond-mat/0412490.*

Wicks, D., Lieberman, E. B., Associates, K., & Division, U. S. F. H. A. T. S. (1980). *Development and Testing of INTRAS, a Microscopic Freeway Simulation Model: Vol. 1, Program Design, Parameter Calibration and Freeway Dynamics Component Development: Final Report*: Federal Highway Administration, Traffic Systems Division.

Wiedemann, R. (1974). Simulation of road traffic flow. *Reports of the Institute for Transport and Communication, University of Karlsruhe.*

Williams, J. (1996). Macroscopic Flow Models in Traffic Flow Theory. *Washington, DC: US Federal Highway Administration*, 6-1.

Williams, J. C. (1997). Macroscopic flow models. *Revised monograph on traffic flow theory.*

Yikai, K., Satoh, J., Itakura, N., Honda, N., & Satoh, A. (1993). *A fuzzy model for behaviour of vehicles to analyze traffic congestion.*

Zhang, L., McHale, G., & Zhang, Y. (2003). Modeling and validating CORSIM freeway origin-destination volumes. *Transportation Research Record: Journal of the Transportation Research Board, 1856*(-1), 135-142.

5

Microwave Beamforming Networks for Intelligent Transportation Systems

Ardavan Rahimian
University of Birmingham
United Kingdom

1. Introduction

An Intelligent Transportation System (ITS) is a system based on wireless communications which has been investigated for many years in order to provide new technologies able to improve safety and efficiency of road transportation with integrated vehicle and road systems. It combines all aspects of technology and systems design concepts in order to develop and improve transportation system of all kinds. ITS, which utilise information and communications technology in vehicle as well as within the roadside infrastructure, can also be used to improve mobility while increasing transport safety, reducing traffic congestion, maximising comfort, and reducing environmental impact (Andrisano et al., 2000). Intelligent transportation systems and applications can improve the quality of travel by selecting routes with up-to-the-minute information data, giving priority to response vehicle teams, notifying drivers about road incidents, and delivering ITS services to drivers. They can reduce fuel consumption by routing the vehicles to their destinations so that fuel waste is significantly reduced, and also fully utilise the capacity of the existing road vehicular networks by controlling the flow of vehicles based on traffic monitoring and detecting congestions.

Vehicles within the ITS framework have to work in an autonomous manner to sense the driving environment and in a cooperative manner to exchange information data such as braking and acceleration between vehicles and also traffic, road, and weather conditions between vehicles and roadside units (Han & Wu, 2011). Hence, radio communications links between vehicles on a motorway are envisaged, leading to the formation of ad-hoc networks between clusters of vehicles and roadside beacons. System performance and analysis can be improved in various ways by the use of smart antenna systems and techniques. These microwave systems fulfil the requirements of improving coverage range, capacity, data-rate, and quality of service (QoS). Smart antenna systems are generally classified as either switched-beam antenna systems or adaptive arrays. Switched-beam antenna systems use fixed multiple beamforming networks (BFNs) in order to create various beam patterns based on the different microwave beamforming techniques and technologies. These smart systems can be used to increase the wireless channel capacity limited by the presence of interference. By using narrow beams available from these systems, it is possible to increase the gain in the desired signal direction and to reduce it toward interference directions. They can also be used for mobile communication base stations in order to provide space-division multiple access (SDMA) capabilities. On the other hand, growing demand in intelligent transportation systems means there is a need for multiple antennas with multiple beams.

Switched-beam antenna systems can greatly improve the performance of the intelligent transportation systems by providing better link quality and high immunity to interference.

Also, generating multiple beams using an array along with having wide bandwidth and beam steering capability are of crucial importance for modern wireless communication systems. For this purpose, various multiple beamforming networks are introduced to have control over the amplitude and phase at each element of the antenna array. Microwave passive networks form an important class of these networks and they have been widely used in switched-beam antenna systems. Two well-known examples of such networks are the Rotman Lens (lens-based beamforming network) and the Butler Matrix (circuit-based beamforming network). They increase the system capacity and provide higher signal-to-interference ratio, consequently enhancing the overall automotive telematics performance.

This chapter presents the novel designs of steerable microwave beamforming networks employing an 8×8 Rotman Lens for operation at 6.3 GHz (C-band), and cascaded 4×4 Butler Matrices for operation at 3.15 GHz (S-band). The microwave beamforming networks are intended for intelligent transportation systems and applications. Although the frequency range likely to be allocated to such systems is 63 GHz, where the short transmission range allows multiple frequency re-use, the microwave networks are frequency scaled models to verify the concept. The objective of this investigation is to develop microwave beamforming networks suitable for a use in vehicle-to-vehicle (V2V) and vehicle-to-infrastructure (V2I) communications. The microwave beamforming networks demonstrate appropriateness to develop well-established designs for systems that can be utilised in ITS applications and framework and vehicular ad-hoc network (VANET) telematics which is the convergence of telecommunications and information processing with application of vehicle tracking.

2. Smart antennas and microwave beamforming techniques

In addition to higher bit-rates and smaller error-rates, microwave beamforming techniques can also be utilised in order to improve the signal-to-noise ratio (SNR) at the receiver and to suppress co-channel interference (CCI) in a multi-user vehicular scenario, thus improving the SINR at the receiver. Using microwave beamforming techniques, the beam patterns of the antenna array can be steered in certain desired directions, whereas undesired directions can be suppressed. Consider an antenna array with N antenna elements, which receives an RF signal from a certain direction. Due to the geometry of the antenna array, the impinging RF signal reaches the individual antenna elements at different time instants, which causes phase shifts between the different received signals. If the direction of the impinging RF signal is known, the phase differences of the RF signals can be compensated by means of phase shifters or delay elements, before the received signals are added up. As a result, the overall antenna pattern of the phase array will exhibit a maximum in the direction of the impinging signal. This principle is called microwave beamforming and is shown in Fig. 1, which is equivalent to a mechanical rotation of the array. In a vehicular communication scenario, transmitted RF signals often propagate via a line-of-sight (LOS) path between transmitter and receiver and via paths that are associated with significant reflectors and diffractors in the environment (such as large trucks). If the directions of these dominant propagation paths are known at the receiver side, microwave beamforming can be applied in order to adjust the receiver beam pattern such that it has a high directivity towards the dominant angles of reception to accomplish significant antenna gains. Due to the required equipment and processing power, however, the use of smart antennas is currently limited to stations that are fixed on vehicles (Mietzner et al., 2009).

Smart antennas are also beneficial in multi-user vehicular scenarios, in order to suppress CCI. Again, both transmitter- and receiver-sided microwave beamforming can be employed for mitigating CCI. When transmitting, each user can adjust the beam pattern such that there are nulls in the directions of other co-channel users and a high directivity towards the desired direction of radiowave transmission. Hence, the SINR for the other co-channel users is improved as well as the SNR at the desired receiver. Similarly, when receiving each user can adjust the beam pattern such that directions of other co-channel interferers are nulled and desired directions of reception are enhanced and therefore each user can improve the received SINR. The use of smart antenna systems for CCI cancellation offers the opportunity to accommodate multiple co-channel users within the same frequency band. This concept is referred to as SDMA (Mietzner et al., 2009).

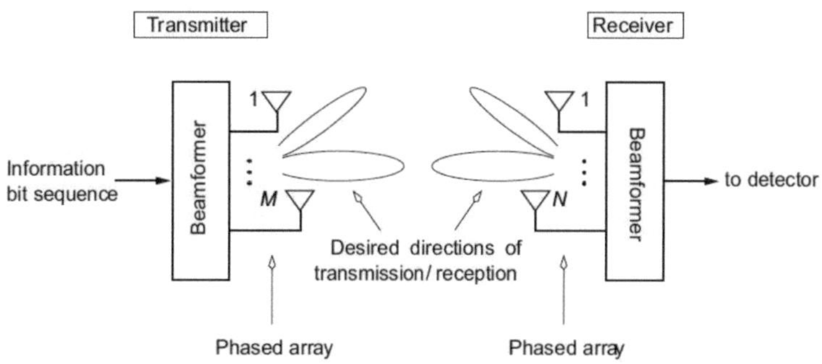

Fig. 1. Rotman Lens Microwave BFN Configuration, taken from (Mietzner et al., 2009).

A multiple microwave beamforming network is one with a capability to form many beams in different directions from the same aperture. If a separate RF transmit or receive system is connected to each beam port, simultaneous independent operation in many directions can be obtained. Alternatively, a single transmit or receive system can be connected to the beam ports through a multiple-way RF switch giving a sequentially scanning antenna.

Switched-beam smart antenna systems may be cheaper than an equivalent phased array, particularly when few beam signals are needed. The creation of a multiple beam antenna using a multiple microwave beamforming network has the advantage that no devices for frequency changing are necessary. The technique therefore has the potential to be simpler and lower in cost than IF, digital, and optical frequency methods. Indeed many antenna configurations, such as lenses, have inherent multiple beam capabilities. In these cases it is only necessary to replace the single feed by an array so that each array element forms one of the multiple beams. The field of microwave beamforming networks encompasses two main research areas namely lens-based quasi-optic types, involving a hybrid arrangement of either a lens objective with a feed array, and circuit-based types used to feed antenna arrays. Circuit-based microwave beamforming networks use transmission lines, connecting power splitters, and hybrid couplers in order to form multiple beam networks. The phase shifts required to produce multi-beam scanning are provided by lengths of transmission line. Lens-based microwave beamforming networks will produce high-gain beams over narrow scan ranges with lenses giving better beam control due to their increased design degrees of freedom. Circuit-based networks (Butler Matrix) have the travelling wave or corporate feed characteristics and can be used in limited size arrays as can the Rotman Lens, which in addition will give wide bandwidth (Hall & Vetterlein, 1990).

3. Rotman lens microwave BFN analysis and design

The Rotman Lens is an attractive passive microwave lens-based beamforming network due to its low cost, reliability, design simplicity, and wide-angle scanning capabilities. It is a device that uses the free-space wavelength of a signal injected into a geometrically configured waveguide to passively shift the phase of inputs into a linear antenna array in order to scan a beam in any desired signal pattern. It has a carefully chosen shape and appropriate length transmission lines in order to produce a wave-front across the output that is phased by the time-delay in the signal transmission.

A Rotman Lens achieves beam scanning using equivalent time-delays that are created by the different path lengths to the radiating elements. These lengths depend on the relative position between the beam port and the array ports on the structure. As long as the path lengths exhibit constant time-delay behaviour over the bandwidth, the lens is insensitive to the beam squint problems exhibited by constant phase beamformers (Weiss, 2010). Each input port will produce a distinct beam that is shifted in angle at the system output. The design of the lens is controlled by a series of equations that set the focal points and array positions. The inputs, during the design of the system, include the desired number of beams and array elements, and the spacing of the elements (Penney, 2008). Since Rotman Lens is a true-time-delay (TTD) device, it produces beam steering independent of frequency and is therefore capable of wide-band operation. Also, the cost of a Rotman Lens implemented on microwave material primarily driven by the cost of the material itself and the price of photo etching (Weiss et al., 2007).

3.1 Rotman lens microwave BFN contour synthesis

The synthesis of the microwave lens assumes several input parameters which are used to compute the inner contour (array contour) point as well as the line lengths. These parameters are element spacing (η), focal ratio (g), lens width (G or F_0), and scan angle (a) (Fig. 2). The lens inner contour points and transmission line lengths are solved for using the technique of path length comparison (Rotman & Turner, 1963).

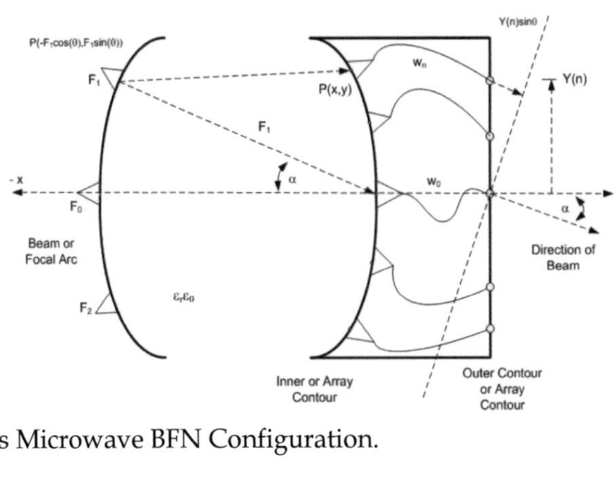

Fig. 2. Rotman Lens Microwave BFN Configuration.

$$F_1P + W(N) + N \sin a = F + W(0) \quad (1)$$

$$F_2P + W(N) - N \sin a = F + W(0) \quad (2)$$

$$F_0P + W(N) = G + W(0) \quad (3)$$

where

$$(F_1P)^2 = (F \cos a + X)^2 + (F \sin a - Y)^2$$

$$(F_2P)^2 = (F \cos a + X)^2 + (F \sin a + Y)^2$$

$$(F_0P)^2 = (G + X)^2 + (Y)^2$$

Lens dimensions are then normalised by the off-axis focal length (F).

$$\eta = N / F \quad (4)$$

$$G = G / F \quad (5)$$

$$x,y = X,Y / F \quad (6)$$

$$w = \frac{W(N)\text{-}W(0)}{F} \quad (7)$$

Manipulation of these equations leads to the following relations for x, y, and w:

$$y = \eta\,(1 - w) \quad (8)$$

$$x^2 + y^2 + 2a_0x = w^2 + b_0^2\eta^2 - 2w \quad (9)$$

$$aw^2 + bw + c = 0 \quad (10)$$

The lens design program solves for these points each time F, η, a, g are modified.

3.2 Rotman lens microwave BFN performance and phase error analysis

In order to calculate performance of the microwave lens, the coupling between ports is approximated using aperture theory and a uniform distribution to the port aperture is implied. These port radiation patterns are used to compute the direct path and reflected path propagation from port to port. Also, to improve the response of the outer beams, the beam and array ports are adjusted so that each line is pointing toward the centre of the lens on the opposite side rather than being normal to the microwave lens surface (Maybell, 1981).

Phase error is calculated by comparing electrical lengths along two distinct paths from a given beam port through the microwave lens (Fig. 3). The first path travels through any one of the off-axis array ports, through its taper and transmission line, and finally along the path from the array element phase centre to the beam phase front. The second path begins at the same beam port but travels through the centre of the array curve and through a length of line common to all array ports. The comparison of these electrical lengths obtains the phase error for this beam port. This is done over the list of beam ports to produce a phase error.

$$\text{Phase Error } \Delta\varphi = R_b - R_a \quad (12)$$

where

$$R_b = |R_b| + \varphi_{TLi} + y(n)\sin(a_i)$$

$$R_a = |R_a| + \varphi_{TL0}$$

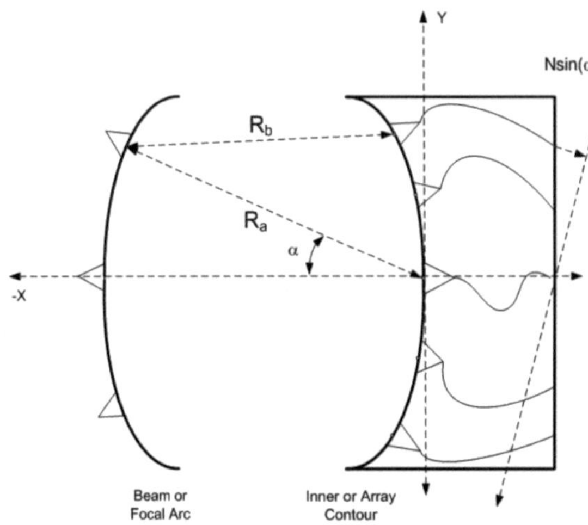

Fig. 3. Rotman Lens Microwave BFN Phase Error Calculation.

When a feed point is placed at any one of the focal points, the corresponding wave-front has no phase error. When the feed is displaced from these lens focal points, the corresponding wave-front will have a phase error. However, for wide-angle microwave beam scanning, the lens must be focused at all intermediate points along the focal arc.

3.3 8×8 ITS Rotman lens microwave BFN design and performance

As Fig. 4 indicates, a realistic 8×8 ITS Rotman Lens microwave beamforming network is designed and simulated for a use in intelligent transportation systems and applications. The design parameters are based on those used in previous section. In this case, the microwave lens is designed to have 8 beam ports, 8 array ports suitable for an 8-element antenna array, a beam scan angle of ±50° at a centre frequency of 6.3 GHz, and an element spacing of 28 mm. The prototype for the lens is fabricated as a microstrip with a 50 Ω impedance transmission lines on Taconic TLC-30 substrate with the dielectric constant (ε_r) of 3.0, substrate thickness (H) of 1.3 mm, and loss tangent of 0.003. The design gives the microwave lens a compact size of 35.91 cm × 25.80 cm. The array ports have also been spaced in such a way that elements of the antenna array can be directly attached to the microwave network. Design, synthesis, and analysis of the 8×8 microwave Rotman Lens and its variants are based on real-time analysis of geometrical optics (GO).

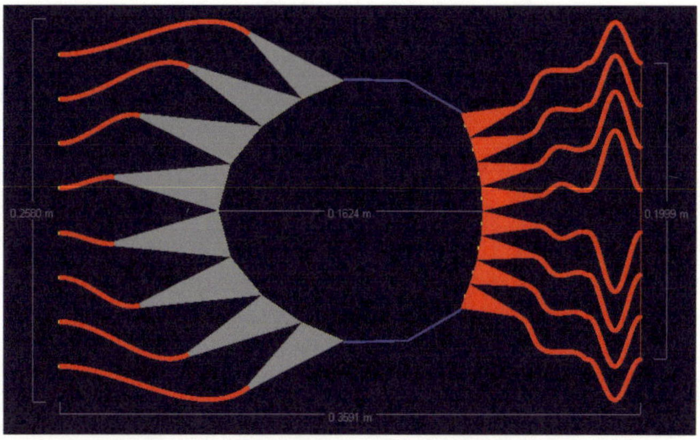

Fig. 4. 8×8 ITS Rotman Lens Microwave BFN Configuration.

The above microwave lens has an elliptical curvature on the beam port side. In this design, dummy ports are replaced with the terminated absorber sidewalls in order to introduce a novelty in the microwave lens structure, reduce the network size and unwanted reflections (and therefore reduce phase errors at the array ports), and increase the performance of the lens. The geometry of the transmission line routing is adjusted in a way to ensure no overlapping, proper spacing between lines, proper curvature, and maintaining overall lens physical length requirement. To obtain the desired performance, the lens requires to be tuned in terms of phase error or the array factor. The tuning involves adjusting the focal ratio (g) of the lens that will minimise the error reported by the phase error. This factor determines the curvature and focus of the lens, and if not adjusted accurately, will produce a messy beam. Hence, the focal ratio (g) is adjusted to 1.2670 in order to minimise the beam to array phase error and to produce well-focused beams (Fig. 5).

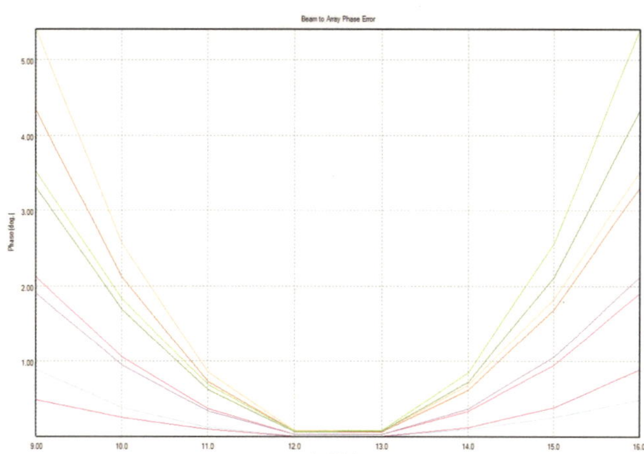

Fig. 5. 8×8 ITS Rotman Lens Microwave BFN Beam to Array Phase Error.

The simulated results for the 8×8 ITS Rotman Lens steerable beamforming network indicate the expected outcomes by having the main lobe of the array factor radiation pattern more than 10 dB greater than the side lobes and having a linear phase shift at each output port as a function of frequency. The beam to array coupling amplitude (the array ports distribution from a given beam or set of beams) for the array ports 9 to 16 has expected outcome of -9 dB to -13 dB has been obtained as a result of the accurate lens design and it confirms how the amplitude distribution along the array contour is much more uniform with beam port pointing enabled. Also, the progressive phase shift for the lens array ports exciting beam port 5 ensures the generation of eight distinct beams and beam scanning capabilities. It is computed using the linear distance between the ports in the dielectric medium chosen.

The fabrication of microwave network was carried out and the lens was then extensively measured on a network analyser over a frequency range of 5.5 GHz to 7.5 GHz with the frequency of operation as 6.3 GHz. Fig. 6 indicates the fabricated microwave network being tested. Only one beam port and one array port (S_{21}) are measured at a time and all other ports are perfectly terminated using 50 Ω termination loads.

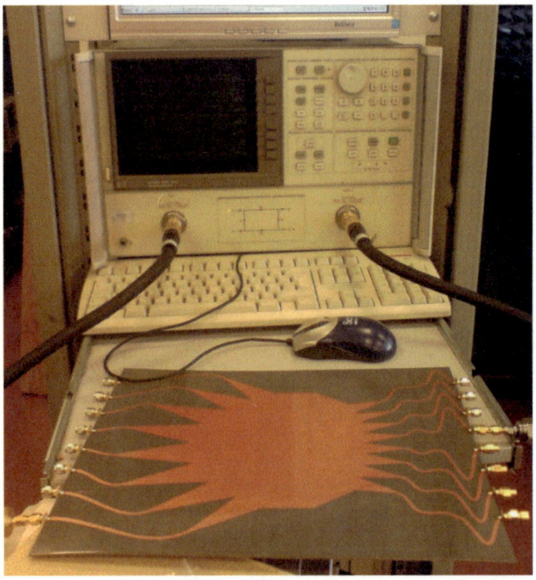

Fig. 6. Fabricated 8×8 ITS Rotman Lens Microwave BFN.

The array factor (*AF*) is a function of the geometry of the antenna array element and the excitation phase and it quantifies the effect of combining radiating elements in an array without the element specific radiation pattern taken into account. The array factor of an *N*-element antenna array is given by (Raisanen & Lehto, 2003):

$$(AF) = \sum_{n=1}^{M} a_n \cos[(2n-1)u]$$
$$where, u = \frac{\pi d}{\lambda} \cos\theta \tag{13}$$

where *d* is the separation between the elements, *M* is the number of isotropic elements, and a_n's are the excitation coefficients of the array elements. By substituting the values to (13), the array factor radiation patterns for the proposed microwave beamforming network can be computed to verify the ITS beam beam scanning electronically steered arrays concept.

As Fig. 7a and Fig. 7b indicate, the array factor radiation patterns for ports 1 to 8 have their beams formed in the expected signal directions having their main lobes showed strong identity with at least 10 dB of isolation from the side lobes. It is shown that despite the non-ideal performance of microwave network, in terms of phase and amplitude distributions, it

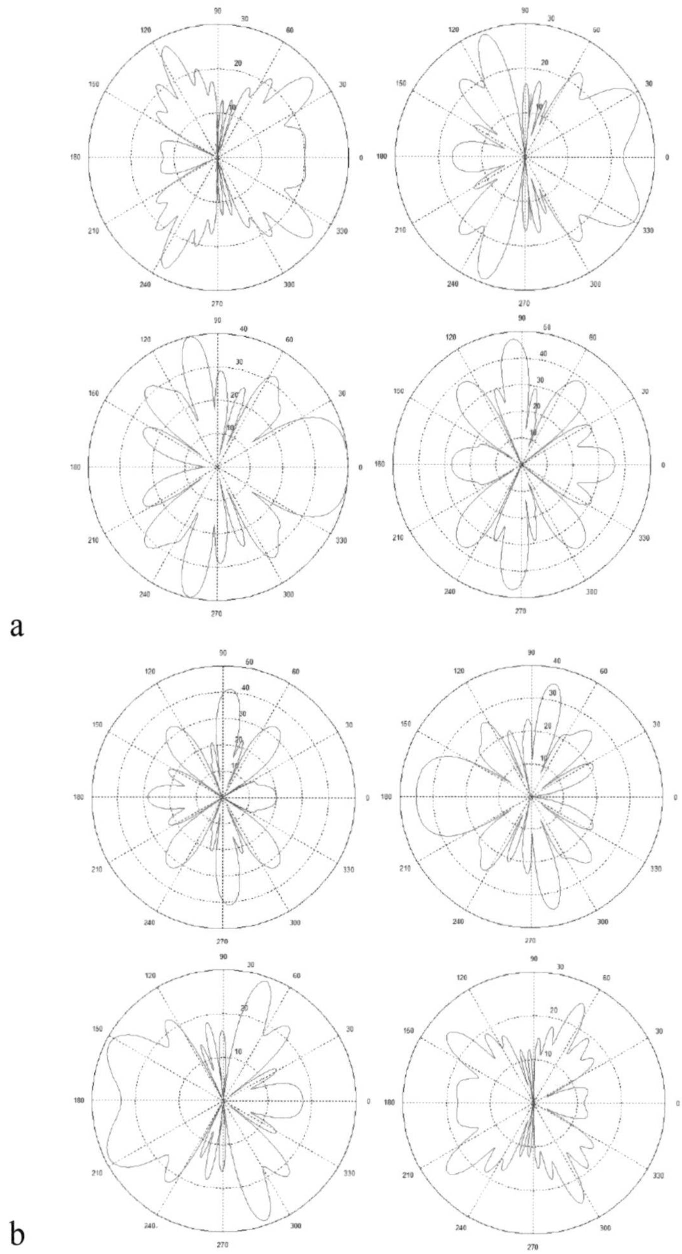

Fig. 7. a. 8×8 ITS Rotman Lens Microwave BFN Measured Array Factor Radiation Patterns for Port 1 to Port 4. b. 8×8 ITS Rotman Lens Microwave BFN Measured Array Factor Radiation Patterns for Port 5 to Port 8.

is still capable of forming well-defined beams suitable for the beam steering experiments by causing the main lobe to be directed in certain directions for ports 1 to 8.

The difference in signal beam shape between the measured radiation patterns and simulated array factors is mainly in the nulls between the beams, which are not deep enough as the measured results because of small phase and amplitude deviations, the cross-coupling effects that are not taken into account in simulations, non-uniformity of transmission line width, and errors occurred during fabrication, measurement, and soldering. By eliminating the mentioned errors and using a shielded metal box with absorbing foams attached to the inner lid to reduce the interferences, the overall system performance will be improved and enhanced in terms of achieving high-gain narrow-beams with desired directions, and the relative phase shift will have a uniform distribution.

The proposed system can be used as the radio zone control technology units which scans the radio zone (antenna beam) in accordance with the average speed of a vehicle group in order to decrease the number of handovers within the specified continuous area. The microwave lens can also be integrated with amplifiers between the lens and the radiating elements as well as an RF switch array for selection of the signal beam ports and an A/D converter which samples the received signal and converts it to a digital signal and a DSP processor unit which then performs a Fast Fourier Transform (FFT) of the digital signal, and the amplitude and the phase parts are separated out.

ITS applications have generally been classified into three main categories with respect to their functionalities as safety, efficiency, and comfort applications. Safety applications minimise the risk of accidents and reduce the severity of the accident if it still occurs (collision avoidance, road sign notifications, incident management). Efficiency applications increase traffic efficiency by managing the traffic flow, and monitoring vehicles and road conditions. The purpose of comfort applications is to provide entertainment facilities and information to passengers by means of Internet access technologies (Dar et al., 2010).

Integration of digital signal processing unit with microwave beamforming network based on the 8×8 Rotman Lens will form a hybrid microwave-digital distributed beamforming network that can be employed in vehicular phased arrays and collision avoidance radar systems in order to support the ITS safety applications. Also, the proposed microwave network system can further be extended to wide-band structures to support the frequency of operation of 63 GHz for vehicle communication systems.

4. Butler matrix microwave BFN analysis and design

The Butler Matrix which is recently used due to its easy fabrication process and low cost, is a method of feeding an antenna array. It requires N beam (input) ports, N output ports, $(N/2)\log_2(N)$ hybrid couplers, and $(N/2)\log_2(N-1)$ fixed phase shifters to form the $N \times N$ network for an N-element array (Ahmad & Seman, 2005). When a signal excites an input beam port of the matrix, it produces different inter-element phase shifts between the output ports. To calculate the number of crossovers needed (14) may be used (Corona & Lancaster, 2003):

$$C_p = 2C_{p-1} + 2^{p-2}(2^{p-2} - 1) \tag{14}$$

where p is the matrix order, which is related to the number of ports by $N=2^p$. In equation (14), p should be equal or greater than 2 and $C_1 = 1$.

In this ITS microwave beamforming network, four hybrid couplers, two crossovers, and two fixed phase shifters are combined to obtain the 4×4 Butler Matrix. The phase differences are ±45° and ±135° from port 1 and port 4, and port 2 and port 3, respectively. The output ports have been spaced in such a way that elements of the antenna array can be directly attached to the microwave network. If the matrix is connected to an antenna array, then the network will act so that the array will have a uniform amplitude distribution and constant phase difference between adjacent elements to generate orthogonal beams. The Butler Matrices are then cascaded in order to produce narrow-beam and broad-beam output that could provide multi-channel operation for automotive telematics applications, particularly for vehicle-to-vehicle (V2V) and vehicle-to-infrastructure (V2I) automotive communications.

The Butler Matrix microwave beamforming network is theoretically lossless in that no power is intentionally dissipated in terminations. There will always be, however, a finite insertion loss due to the inherent losses in the hybrid couplers, fixed phase shifters, and transmission lines that make up the matrix. The Butler Matrix passive beamforming antenna also requires that the individual beam patterns be orthogonal in space (Skolnik, 2000).

Independent orthogonal beams mean that when two or more beam input ports are simultaneously excited, the resulting radiation is a linear superposition of the radiations that would be obtained when the ports are excited separately. In addition, when a signal is applied to one port it should have no output at the other ports. An antenna which is lossless and passive means that the radiated power is the same as the input power. Fig. 8 shows the topology of the Butler Matrix. The phase shift at the matrix output ports can be determined by summing up all the phase shifts of signal paths. Table 1 also indicates the resulting phase shift's characteristics at the matrix output elements. It was designed in such a way that when current excited to any input ports will only has one constant as shown in Table 1.

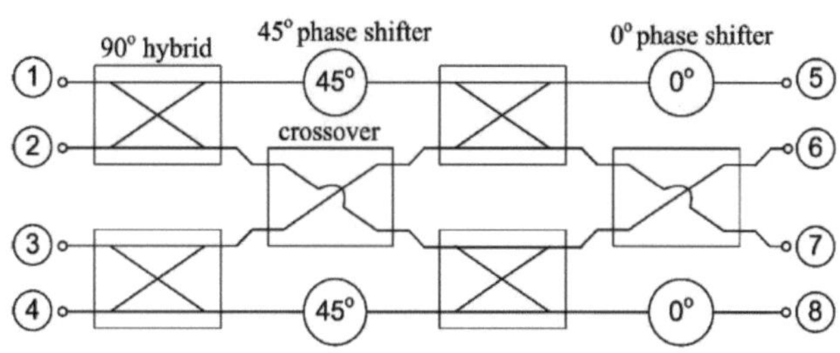

Fig. 8. Topology and Routing of Signal Paths of the 4×4 Butler Matrix Microwave BFN.

Port	*Input Port 1*	*Input Port 2*	*Input Port 3*	*Input Port 4*
Output Port 5	-45°	-135°	-90°	-180°
Output Port 6	-90°	0°	-225°	-135°
Output Port 7	-135°	-225°	0°	-90°
Output Port 8	-180°	-90°	-135°	-45°
Δ Phase	-45°	+135°	-135°	+45°

Table 1. Phase Shift's Characteristics at the Output Ports of the 4×4 Butler Matrix BFN.

Fig. 9 shows the proposed planar configuration of the 4×4 ITS Butler Matrix microwave beamforming network obtained as a result of the accurate branch line coupler, crossover, and phase shifter components design. The input and output ports are connected through the phase shifters and branch line couplers such that when a signal is applied to any input port, the matrix produces equal amplitude signals at all the output ports. The 45 degree and 0 degree phase shifters, together with phase adjustment, are obtained by connecting a transmission line at the output port of the hybrid coupler to the input port of the other hybrid coupler. At the Butler Matrix output ports, additional transmission lines are placed in such a way that antenna array elements can be directly connected to the network. The design gives the 4×4 ITS Butler Matrix network a compact size of 11.6 cm × 9.1 cm for enhanced operation and better performance (Rahimian & Rahimian, 2010).

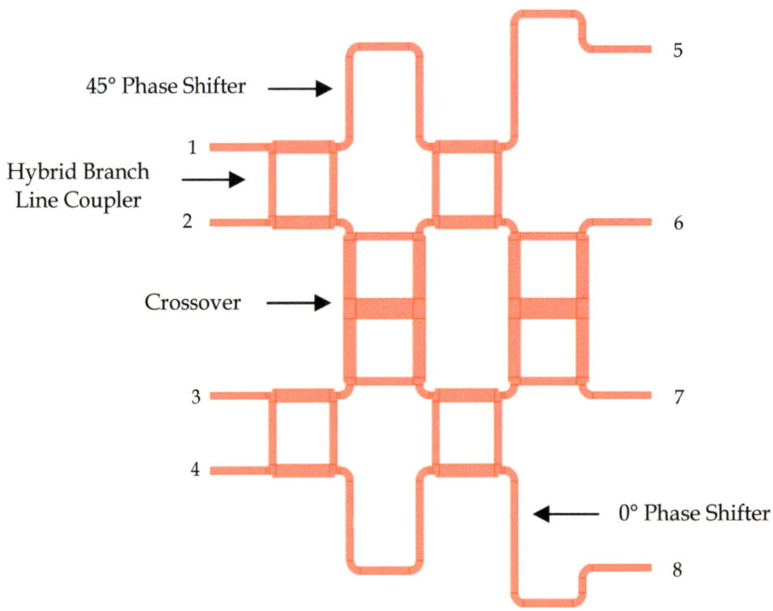

Fig. 9. Layout of the 4×4 ITS Butler Matrix Microwave BFN with Components.

4.1 4×4 ITS Butler matrix microwave BFN realisation

The fabrication of the 4×4 ITS Butler Matrix microwave beamforming network has been carried out and the measured results have slight error compared to simulated results. The prototype for the matrix is fabricated as a microstrip with a 50 Ω impedance transmission lines on FR4 substrate with the dielectric constant (ε_r) of 4.7 and thickness (H) of 0.8 mm, and loss tangent of 0.01. The ITS Butler Matrix microwave beamforming network was then extensively measured on a network analyser over a frequency range of 2.5 GHz to 3.5 GHz with the frequency of operation as 3.15 GHz. The Butler Matrix has been shielded with a metal box along with absorbing foam attached to the top lid of the inner box in order to reduce internal coupling and external interference effects. Fig. 10 indicates the fabricated ITS beamforming network being tested. Only one beam port and one output port are measured at a time and all other ports are perfectly terminated using 50 Ω termination loads.

Fig. 11 and Fig. 13 indicate the simulated and measured S-parameters exciting beam port 1 and beam port 2 respectively. At an operating frequency of 3.15 GHz, the simulated results agree with the measured results. As Fig. 12 and Fig. 14 indicate, the measured progressive phase shift for the Butler Matrix microwave beamforming network output ports exciting port 1 and port 2 respectively ensure the generation of four different beams at 3.15 GHz.

By using the narrow-beam signals available from the ITS Butler Matrix, it is possible for a vehicle to increase gain in the desired signal directions and reduce the gain in interference signal directions. The differences between the simulations and measurements and slight distortion of the beam shape might be due to the employed FR4 board and fabrication process, non-uniformity of matrix transmission line width, cross-coupling effects, and measurement errors. This Butler Matrix microwave network can be used as a planar passive BFN for multi-beam antennas used in automotive telematics and ITS applications.

The microwave beamforming network was designed to be placed anywhere on the envelope of the symmetrical cut-plane running through the centre of the vehicle. Possible antenna placement positions are therefore on the roof of the vehicle or inside the front and rear bumpers with a plastic radome. Antenna location is important to permit a mounting, which has little impact on vehicle styling, be of low cost, and be capable of addition to the vehicle with minimum re-design of surrounding components.

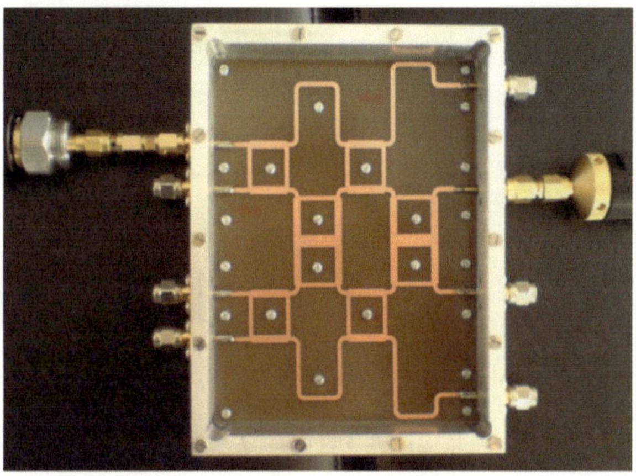

Fig. 10. Fabricated 4×4 ITS Butler Matrix Microwave BFN.

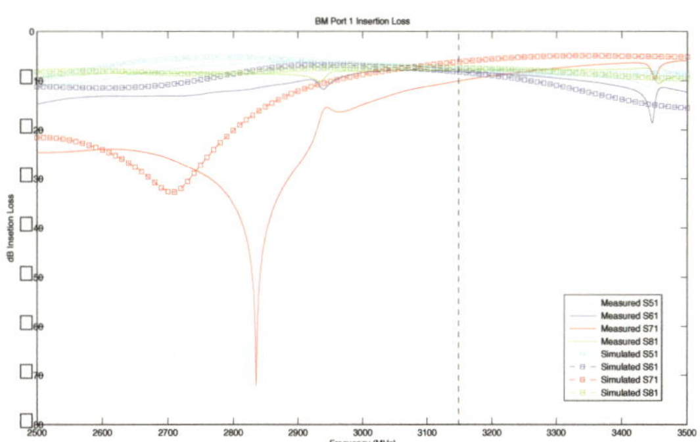

Fig. 11. 4×4 ITS Butler Matrix Microwave BFN Simulated and Measured S-Parameters Exciting Beam Port 1.

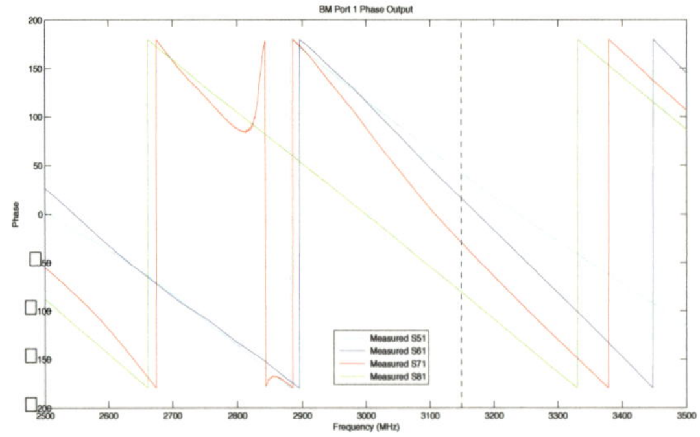

Fig. 12. 4×4 ITS Butler Matrix Microwave BFN Measured Phase Shift for Output Ports Exciting Beam Port 1.

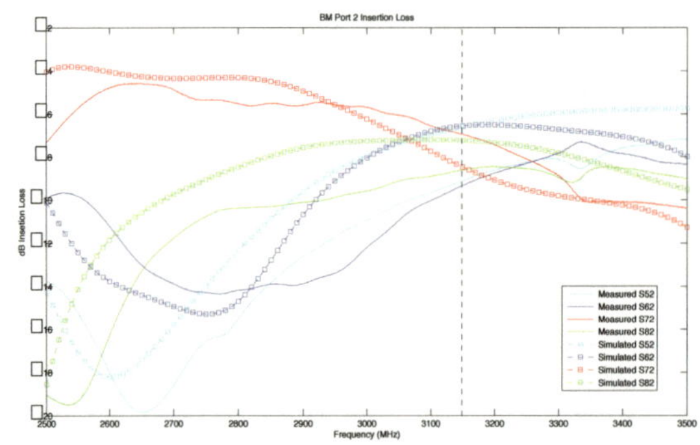

Fig. 13. 4×4 ITS Butler Matrix Microwave BFN Simulated and Measured S-Parameters Exciting Beam Port 2.

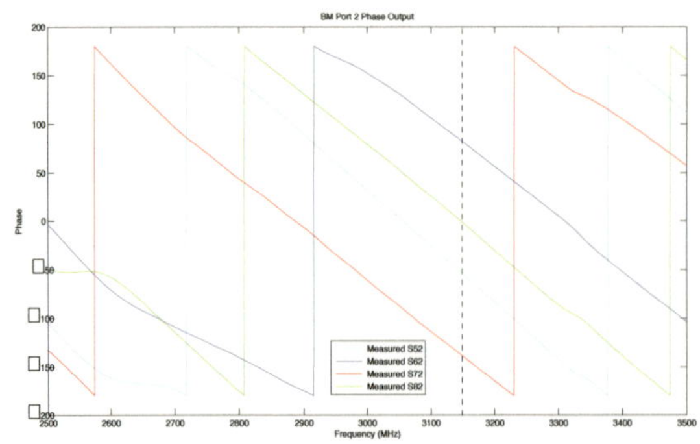

Fig. 14. 4×4 ITS Butler Matrix Microwave BFN Measured Phase Shift for Output Ports Exciting Beam Port 2.

4.2 Cascaded ITS Butler matrices microwave BFN design and performance

As Fig. 15 indicates, the proposed ITS Butler Matrix microwave beamforming network has been cascaded back-to-back in order to produce narrow-beams and broad-beams suitable for V2V and V2I automotive communications. Signals entering the input ports of the first Butler Matrix microwave beamforming network are subdivided into equal amplitude with progressive phase variation across the matrix output ports, for high-gain and narrow-beam reception that are potential for ITS long-range automotive communication.

These signals are then fed into the Wilkinson power dividers. The signal from one end of each Wilkinson power divider forms a narrow output beam while the signals from the other ends of the Wilkinson power divider are fed into the second Butler Matrix network in order to regenerate the broad-beam signal characteristics of the individual radiating elements. As a result, high-gain and narrow-beam signals are produced on the output of the first Butler Matrix network while broad-beams are produced on the output ports of the second Butler Matrix network that are suitable for ITS short-range automotive communication.

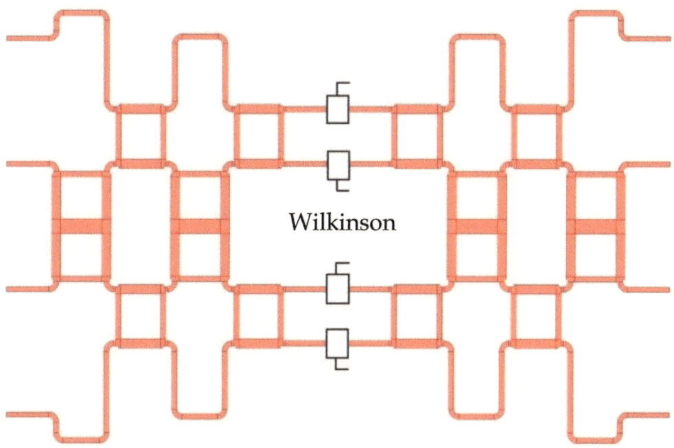

Fig. 15. Block Diagram of the Cascaded ITS Butler Matrices Microwave BFN.

Fig. 16 and Fig. 17 present the computed array factor radiation patterns suitable for ITS long-range application and short-range application respectively. The concept of cascaded ITS Butler Matrices microwave beamforming network has been examined in which the first Butler Matrix microwave beamforming network will act as a power divider and the second Butler Matrix network will act as a combiner in order to produce high-gain narrow-beams for long-range communication from the outputs of the first Butler Matrix beamforming network and broad-beam signals for short-range communication from the outputs of the second Butler Matrix microwave network.

The ITS microwave BFN system can further be integrated with low-noise amplifiers (LNAs) to increase the gain and to reduce the noise power as well as an RF switch array for selection of the input beam ports of the network. A control circuit switches the RF switch to switch the oscillator signal rapidly among beam ports by changing the feed points at a specified rate. At the array ports, the phase shifted signals are amplified via an amplifier and then radiated through the antennas. On the receiving side, the receiver amplifies the signal and then a bank of filters filter the received signal which in turn is fed to the array ports of the network via an RF switch.

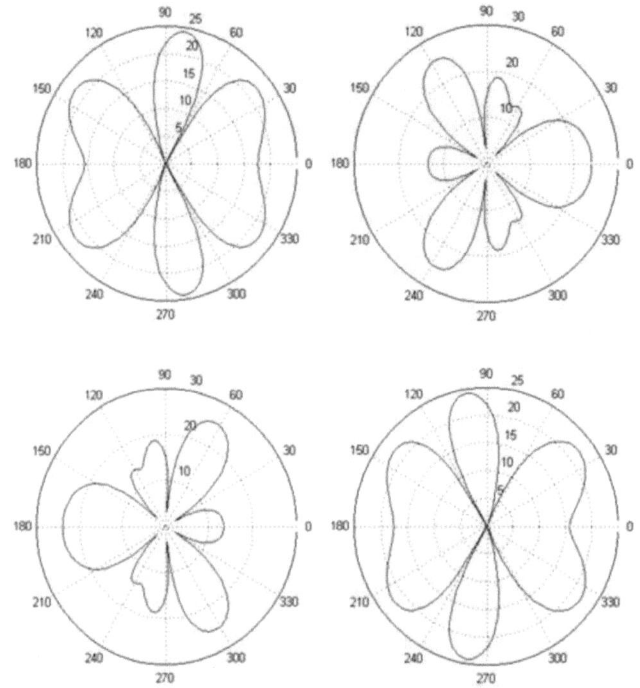

Fig. 16. Cascaded Butler Matrices Computed Narrow-Beam Array Factor Radiation Patterns.

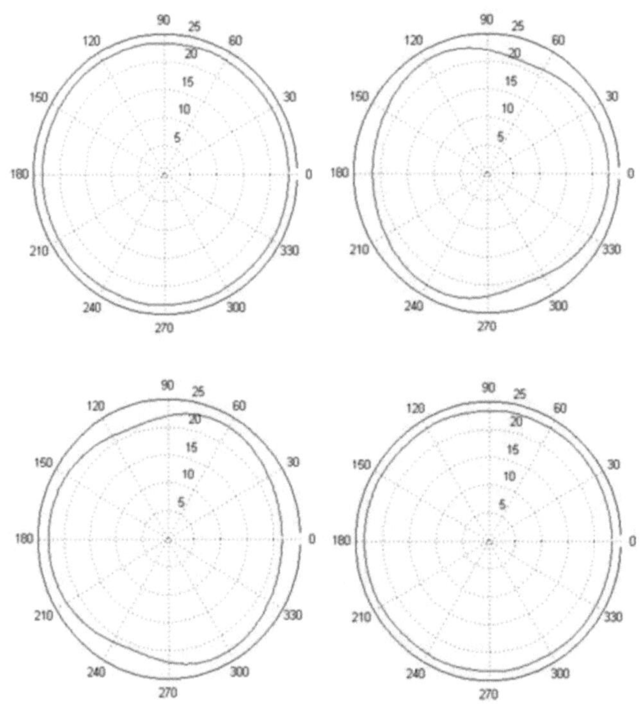

Fig. 17. Cascaded Butler Matrices Computed Broad-Beam Array Factor Radiation Patterns.

5. ITS V2V and V2I automotive communications scenario

ITS is the application of high enabling technology to adaptive traffic signal systems control, congestion charging, information provision, and transit management systems in order to increase and enhance the safety and efficiency of the surface transportation system using radiowave beacons and real-time traffic information communication with major areas as: Multi-modal travel management and traveller information, commercial vehicle operations to achieve safe and cost-effective operation through cooperation and advanced automated networking technologies, and advanced vehicle control and safety systems.

Vehicular telematics are a key technological component of future intelligent road networks. Such systems and technologies offer increased road efficiency, increased safety, improved communications and information services to drivers and passengers, and reduced road congestion and accident rates. Vehicle-to-infrastracture (V2I) or vehicle-to-vehicle (V2V) communication links are likely to be key elements (Fig. 18). Implementation of intelligent transportation systems and applications and vehicular telematics will require demonstration of a number of microwave systems and technologies at an acceptable price per unit. These technologies include antenna arrays, microwave beamforming networks, transmit/receive components, and a variety of sensors, both road and vehicle mounted, in order to increase road efficiency and provide additional services to drivers and travellers. The antennas and beamforming networks are required to provide steered and switched-beam smart radiation patterns to maintain links to moving vehicles and to compensate for signal fading in a complex and dynamic multipath environment. Hence, the development that will be a key to the provision of information-rich and high data-rate services will be microwave systems capable of providing communication links either with roadside beacons (V2I) or with other vehicles (V2V). In the latter case, it will be possible to form wireless vehicular ad-hoc networks (VANETs) with the benefit of reducing communication link range in high traffic density and providing multiple routes between vehicles and roadside beacons.

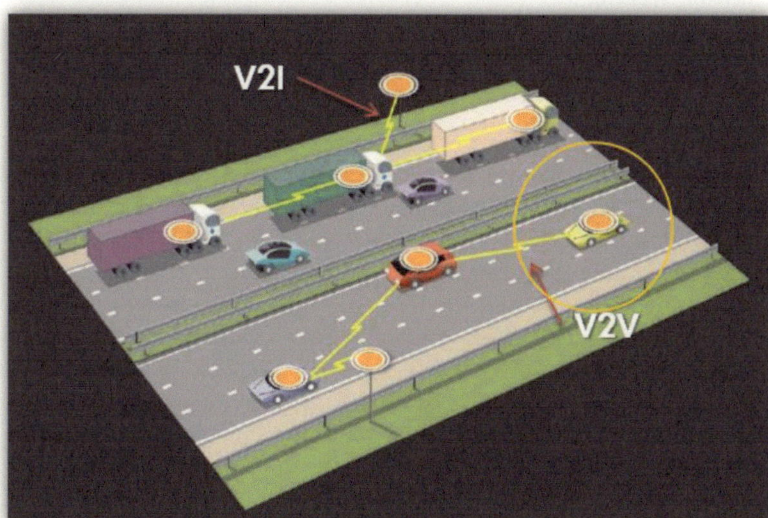

Fig. 18. ITS Concept of V2V and V2I Automotive Communications.

Intelligent transportation systems also play an important role in the research activities on road safety, allowing vehicles to detect a safety hazard and to react to it timely. Through immediate forwarding of hazard warning information to other vehicles via wireless vehicle-to-vehicle (V2V) communication, other vehicles could avoid running into the hazardous situation. The same wireless communication interface could be used to provide the vehicle with traffic control and road safety information from roadside infrastructure via wireless vehicle-to-infrastructure (V2I) communication. Both V2V and V2I are the basis for ITS framework and applications providing a potential for avoidance of accidents.

In order to ensure efficient allocation of RF resources, it is important to group various V2V- and V2I-based applications in different categories based on their need for radio resources. The first such category is the ITS V2V-based Critical Road Safety Applications characterised by strict time constraints where one vehicle must warn another vehicle of a sudden safety hazard instantaneously. Such ITS applications have strict requirements on communication reliability, tolerable transmission latency, minimum throughput, and medium access delays.

Second category is the ITS V2I based Safety and Traffic Efficiency Applications which are informational applications. These applications may be less time-critical and may benefit from central resource management by roadside infrastructures, more RF link stability due to roadside unit's static nature, and better antennas. Depending on their unique requirements, Critical Road Safety Applications and Safety and Traffic Efficiency Applications require a higher QoS, such as instant access to RF frequency channel, high SNR and low channel interference, and reliable wireless communication to ensure that the safety messages are received by vehicles with high probability and for both types of applications, microwave beamforming networks based on Butler Matrices and Rotman Lenses can be employed in order to establish a reliable RF communication link among vehicular beacons.

A study was also carried out of Medium Access Control (MAC) protocols which are suitable for V2V and V2I automotive communications. The aim is to be able to communicate within a group of vehicles travelling as a cluster, between vehicles and a roadside transceiver, and from a roadside transceiver in a broadcast mode. Telematics architectures available in vehicular communication and networks are Vehicle Infrastructure Integration (VII) and Communication Access for Land Mobiles (CALM).

VII architecture seeks significant improvement in vehicle safety, mobility, and commerce by deploying a communication infrastructure on roadways and installing Dedicated Short Range Communication (DSRC) radios on all production light vehicles (Fig. 19). In this scenario, Onboard Unit (OBU) is located inside vehicle, Roadside Unit (RSU) is located on the road and acts as a data gathering and distribution point, control channel broadcasts application and vehicular communication establishment, and service control establishes communication between OBUs and RSUs and between OBUs. Also, DSRC which is a short to medium range communications service that supports both public safety and private operations in V2I and V2V communications is meant to provide very high data transfer rates for mobile wireless nodes in relatively small communication zones and with small latency.

Wireless Access in Vehicular Environment (WAVE) is the mode of operation used by IEEE 802.11 devices in the DSRC band allocated for ITS communication. Fig. 20 shows the WAVE system components. All these advanced wireless vehicular ad-hoc networks (VANETs) can further be integrated with advanced distributed microwave beamforming networks in order to form a state of the art vehicular network with enhanced performance to serve the ITS framework and objectives. The inherent capabilities of microwave beamforming networks and techniques together with VANETs complex algorithms and architectures will provide a powerful synergy for intelligent transportation systems and vehicular telematics realisation.

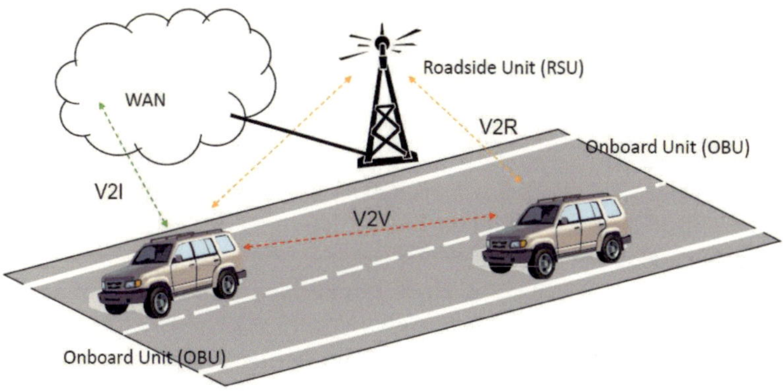

Fig. 19. ITS Vehicle Infrastructure Integration Architecture Equipment Types.

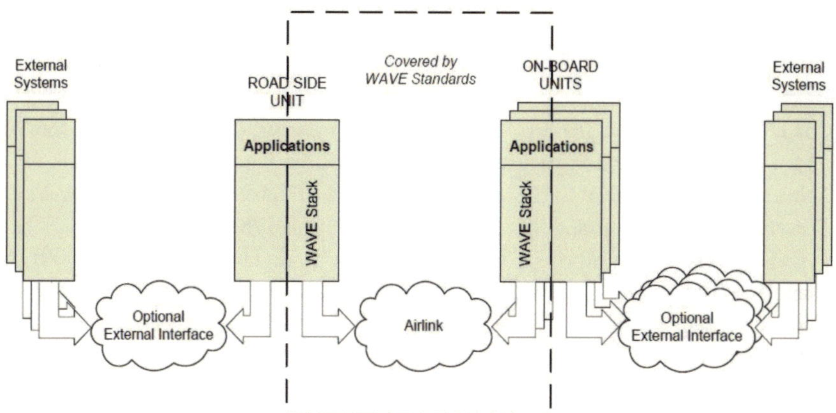

Fig. 20. ITS Wireless Access in Vehicular Environment (WAVE) System Components.

6. Conclusion

The need to relieve traffic congestion and make more efficient use of motorway networks requires a more sophisticated approach to traffic and transportation management. ITS applications and vehicular networks and telematics can offer many benefits using advanced RF and microwave technologies, where vehicles mounted systems communicate with other vehicles or with an infrastructure of roadside beacons. Hence, researches on intelligent transportation systems and applications were carried out to enhance safety and efficiency of road transportation related to vehicle-to-vehicle (V2V) and vehicle-to-infrastructure (V2I) automotive communications. Microwave beamforming networks can greatly increase and enhance the performance of wireless systems used in intelligent transportation systems and framework. In this contribution, passive planar steerable microwave beamforming networks based on Rotman Lens and cascaded Butler Matrices have been designed and analysed in order to support the wireless systems used in vehicular networks, intelligent transportation systems, and collision avoidance program which includes rear-end collision avoidance system, intelligent adaptive cruise control, road departure collision avoidance system, and lane change collision avoidance system.

7. References

Ahmad, S. R. & Seman, F. C. (2005). 4-Port Butler Matrix for Switched Multibeam Antenna Array, *Proceedings of IEEE Asia-Pacific Conference on Applied Electromagnetics*, pp. 69-73, December 2005

Andrisano, O.; Verdone, R. & Nakagawa, M. (2000). Intelligent Transportation Systems: The Role of Third-Generation Mobile Radio Networks. *IEEE Communications Magazine*, (September 2000), pp. 144-151

Corona, A. & Lancaster, M. J. (2003). A High-Temperature Superconducting Butler Matrix. *IEEE Trans. on Applied Superconductivity*, Vol. 13, (December 2003), pp. 3867-3872

Hall, P. S. & Vetterlein, S. J. (1990). Review of radio frequency beamforming techniques for scanned and multiple beam antennas. *IEE Proceedings*, Vol. 137, No. 5, (October 1990), pp. 293-303

Han, L. & Wu, K. (2011). Multifunctional Transceiver for Future Intelligent Transportation Systems. *IEEE Transactions on Microwave Theory and Techniques*, Vol. 59, No. 7, (July 2011), pp. 1879-1892

Dar, K.; Bakhouya, M.; Gaber, J. & Wack, M. (2010). Wireless Communication Technologies for ITS Applications. *IEEE Communications Magazine*, (May 2010), pp. 156-162

Maybell, M. (1981). Ray Structure Method for Coupling Coefficient Analysis of the Two Dimensional Rotman Lens, *Proceedings of IEEE Antennas and Propagation Society International Symposium*, Vol. 19, pp. 144-147, June 1981

Mietzner, J.; Schober, R.; Lampe, L.; Gerstacker, W. H. & Hoeher, P. A. (2009). Multiple-Antenna Techniques for Wireless Communications – A Comprehensive Literature Survey. *IEEE Communications Surveys & Tutorials*, Vol. 11, No. 2, (Second Quarter 2009), pp. 87-105

Penney, C. (2008). Rotman Lens Design and Simulation in Software. *IEEE Microwave Magazine*, Vol. 9, (December 2008), pp. 138-139

Rahimian, A. & Rahimian A. (2010). Enhanced RF Steerable Beamforming Networks based on Butler Matrix and Rotman Lens for ITS Applications, *Proceedings of IEEE Region 8 International Conference on Computational Techniques in Electrical and Electronics Engineering*, pp. 567-572, July 2010

Raisanen, A. V. & Lehto, A. (2003). *Radio Engineering for Wireless Communication and Sensor Applications*, Artech House, Inc., Norwood, MA

Rotman, W. & Turner, R. (1963). Wide-Angle Microwave Lens for Line Source Applications. *IEEE Trans. on Antennas and Propagation*, Vol. 11, (November 1963), pp. 623-632

Skolnik, M. (2000). *Introduction to Radar Systems*, 3rd Edition, McGraw-Hill

Weiss, S. (2010). Low Profile Arrays with Integrated Beamformers. *Army Research Lab*, 2010

Weiss, S.; Keller, S. & Ly, C. (2007). Development of Simple Affordable Beamformers for Army Platforms. *The Army Research Lab*, 2007

6

Deploying Wireless Sensor Devices in Intelligent Transportation System Applications

Kirusnapillai Selvarajah, Budiman Arief, Alan Tully and Phil Blythe
Newcastle University
United Kingdom

1. Introduction

A recent study by the UK Government's Office of Science and Innovation, which examined how future intelligent infrastructure would evolve to support transportation over the next 50 years looked at a range of new technologies, systems and services that may emerge over that period (UK DfT, 2006). One key class of technology that was identified as having a significant role in delivering future intelligence to the transport sector was wireless sensor networks and in particular the fusion of fixed and mobile networks to help deliver a safe, sustainable and robust future transportation system based on the better collection of data, its processing and dissemination and the intelligent use of the data in a fully connected environment. The important innovations in wireless and digital electronics are beginning to support many applications in the areas of safety, environmental and emissions control, driving assistance, diagnostics and maintenance in the transport domain. The last few years have seen the emergence of many new technologies that can potentially have major impacts on Intelligent Transportation Systems (ITS) (Tully, 2006).

U.S. DOT recently launched a 5 Years ITS strategic research plan to explore the potentially transformative capabilities of wireless technology to make surface transportation safer, smarter and greener and ultimately enhance livability for Americans (US DOT, 2011). This research program formerly known as *IntelliDriveSM* and now renamed as "Connected Vehicle Research" program which focus to develop a networked environment supporting very high speed transactions among vehicles (V2V) and between vehicles and infrastructure components (V2I) or hand held devices (V2D) to enable numerous safety and mobility applications (US DOT, 2010).

The European Telecommunications Standards Institute (ETSI) has been creating and maintaining standards and specifications for Intelligent Transportation Systems (ITS) include telematics and all types of communications in vehicles, between vehicles (e.g. vehicle-to-vehicle), and between vehicles and fixed locations (e.g. vehicle-to-infrastructure). The ETSI not only looking for road transport domain but also the use of information and communication technologies (ICT) for rail, water and air transport, including navigation systems (ETSI, 2009).

As future intelligent infrastructure will bring together and connect individuals, vehicles and infrastructure through wireless communications, it is critical that robust communication technologies are developed. Mobile wireless sensor networks are self-organising mobile networks where nodes exchange data without the need for an underlying infrastructure. In the road transport domain, schemes which are fully infrastructure-less and those which use a combination of fixed (infrastructure) devices and mobile devices fitted to vehicles and other moving objects are of significant interest to the ITS community as they have the potential to deliver a 'connected environment' where individuals, vehicles and infrastructure can co-exist and cooperate, thus delivering more knowledge about the transport environment, the state of the network and who indeed is travelling or wishes to travel. This may offer benefits in terms of real-time management, optimisation of transportation systems, intelligent design and the use of such systems for innovative road charging and possibly carbon trading schemes as well as through the CVHS (Cooperative Vehicle and Highway Systems) for safety and control applications.

Within the vehicle, the devices may provide wireless connection to various information and communications technologies components in the vehicle and connect with sensors and other devices within the engine management system. There is growing consumer demand for wireless communication technologies in transporation applications from point-to-point to multiplexed communications. Advances in portable devices (Smartphone, Personal Digital Assistant and Navigation Systems) may exploit the possibility of interconnection using in-vehicle communications. Also advances in wireless sensor networking techniques which offer tiny, low power and MEMS (Micro Electro Mechanical Systems) integrated devices for sensing and networking will exploit the possibility of vehicle to vehicle and vehicle to infrastructure communications (Blythe, 2006).

Wireless sensor networks offer an attractive choice for low-cost and easy-to-deploy solutions for intelligent transportation applications. Intelligent parking lot application using Wireless Sensor Networks is presented in (Lee et al, 2008). In this paper, authors argue the use of both ultrasonic and magnetic sensors in accurate and reliable detection of vehicles in parking lots. A wireless sensor networks scenario for Intelligent Transportation Systems (ITS) is studied in (Chen et al, 2009) and authors also propose new routing protocol to make the WSN more energy efficient and with less delivery latency.

Wireless Sensor Network based adaptive vehicle navigation in multi-hop relay WiMAX networks is proposed in (Chang et al, 2008). This paper proposes the WiMAX multi-hop relay networks as the inter-vehicle communications to increase the reliability and efficiency of inter-vehicle communications. Real-time traffic information is gathered here from various types of sensors equipped on vehicles and exchanged among neighbor vehicles. Collaboration-based hybrid vehicular sensor network architecture is proposed in (Kong & Tan, 2008). In this paper, authors propose a collaborative hybrid method to deliver desired data to particular drivers effectively. The road side sensors and vehicular mobile sensors are used to restore data and exchange data.

As the wireless sensor networks technology is still relatively new and very little is known about its real application in the transport domain. Our involvement in the transport-related projects provides us with an opportunity to carry out research and development of wireless sensor network applications in transport systems. This chapter outlines our experience in the ASTRA (ASTRA, 2005), TRACKSS (TRACKSS, 2007) and EMMA (EMMA, 2007) projects and provides an illustration of the important role that the wireless sensor technology can play in future ITS. This chapter also presents encouraging results obtained from the experiments in investigating the feasibility of utilising wireless sensor networks in vehicle and vehicle to infrastructure communication in real ITS applications.

2. Communication technologies

The use of networks for communications between the electronic control units of a vehicle in production cars dated from the beginning of the 1990s. The Controller Area Network (CAN) was first introduced by BOSCH with the clear intent to serve communication systems for automotive applications and it is still dominant in automotive networks. The CAN is not fully satisfying requirements such as predictability, performance and dependability which are mandatory in automotive communications. To overcome the limitations of the CAN technology, a number of technologies have been developed for designing automotive networks such as Time-Triggered Protocol (TTP), Time-Triggered CAN (TTCAN), Byteflight and Flexray (Navet et al, 2005).

Wireless communication technologies such as ZigBee (ZigBee Alliance, 2006), Bluetooth (Bluetooth, 2006) and Wi-Fi are also expected to be widely employed in the near future in automotive communication. It is evident that wireless communications can be used in-vehicle, inter-vehicle and between vehicle and infrastructure in transportation applications (Cai & Lin, 2008; Kosch et al, 2009; Heddebaut, 2004). Bluetooth is currently the most widely used automotive wireless technology for in-vehicle communication and Wi-Fi is used for vehicle to vehicle communication by several pilot research projects, e.g., the Car2Car consortium (Car2Car consortium, 2007). Ultra Wide Band (UWB) is an emerging wireless technology that uses a very large bandwidth (Yang and Giannakis, 2004). It is targeted for multimedia networking whereas 802.11 networks address data networking. Intelligent collision avoidance and cruise control systems can be developed using UWB technology as those systems need high ranging accuracy and target differentiation capabilities. UWB technology can also be integrated into vehicle entertainment systems by downloading high-rate data from road-side infrastructure UWB transmitters. Communication Air-interface, Long and Medium range (CALM) (ISO TC204 Working Group, 2007) has many potential applications in V2V and V2I communication.

ZigBee will be able to fill the gap left by these other technologies, mainly in the interconnection of wireless sensor devices with vehicles and infrastructure. The ZigBee standard has evolved since its original release in 2004 and it is a low cost low power wireless networking standard for sensors and control devices. ZigBee provides network speeds of up to 250kbps and is expected to be largely used in typical wireless sensor network applications where high data rates are not required. Table 1 shows a comparison between five technologies relating to the most important factors which need to be considered in the ITS application domain.

ZigBee, Bluetooth, Wi-Fi and UWB have been designed for short-range wireless applications with low power solutions and can be used at the in-vehicle and vehicle to infrastructure communication. The ZigBee technology can accommodate larger numbers of devices than Bluetooth. On the other hand, Bluetooth offers high bandwidth with relatively high throughput. Many wireless sensor network applications do not require high data rate communication technology as it is based on data exchange. ZigBee provides 250 kbps data rate and is expected to be enough for the many wireless sensor network applications. Notably, ZigBee uses low overhead data transmission and requires low system resources which are vitally important factors for embedded sensor networks. Also mesh networking features in ZigBee protocol allow devices to extend its coverage and optimize its radio resources. The features show that ZigBee is a suitable communication technology for the wireless sensor networks based ITS applications.

Standard	ZigBee	Bluetooth	Wi-Fi	UWB	CALM
Automotive application	In-vehicle and vehicle to infrastructure	In-vehicle and device connectivity	Vehicle to vehicle and vehicle to infrastructure	In-vehicle, vehicle to vehicle and vehicle to infrastructure	Vehicle to vehicle and vehicle to infrastructure
Network range	Up to 100m	Up to 70m	Up to 100m	Up to 20m	Up to 1000m
Network method	Mesh	P2P	P2P	P2P	P2P
Bandwidth	250 kbps	12Mbps	54Mbps	1000Mbps	54Mbps
Frequency	2.4 GHz	2.4 GHz	2.4 GHz	3.1 GHz	5.8GHz
Advantages	Low power Many devices Low overhead	Dominating PAN Easy synchronisation	Dominating WLAN Widely available	Robust and high bandwidth	Continues communication Wide coverage
Disadvantages	Low bandwidth	Consume medium power	Consume high power	Interference Short range	Consume high power

Table 1. Comparison of Wireless Communication Technologies.

3. Smartdust in transportation applications

Smartdust (or mote) is a new concept for wireless sensor networks which offers tiny, low power and MEMS integrated devices for sensing and networking and it's also known as "motes". It is not only interesting for the low power sensing technologies but also the low power communication and networking capability which it has demonstrated (Tully, 2006). Fundamentally, it provides a convenient and economic means of gathering and disseminating environmental and other useful information in the transport domain. Existence of ZigBee based networking capability between Smartdust and other devices will benefit many low cost and low power applications in the ITS. Smartdust devices also have sensors attached to them to monitor the physical environment in some way. These sensors can be built directly onto the Smartdust or can come as daughter-boards which can be connected in some way to the motes main motherboard. Sensors can measure a wide range of environmental parameters, such as pollution, noise, temperature, humidity as well as vehicle speed, vehicle direction and vehicle presence. Initial studies suggest environmental monitoring, vehicle to vehicle, vehicle to infrastructure and infrastructure to infrastructure applications may exist for Smartdust in the transport domain. Indeed the vital application of the devices is beginning to be tested in the road vehicle environment. Even though a range of Smartdust platforms are available in the market, Crossbow Mica (Crossbow, 2006) family motes will be discussed here due to their commercial success in many wireless sensor network applications.

Mica family motes can be programmed with NesC based TinyOS (TinyOS, 2006) , C based NanoQplus (NanoQplus, 2008) and many other operating systems which are designed for embedded systems with very limited resources. Also, MicaZ mote will be the most suitable platform for the wireless sensor networks based ITS application since it features sensing and networking capabilities with low power consumption using ZigBee as communication protocol. Figure 1 shows a Crossbow MicaZ mote. The MicaZ mote is a family of the Crossbow Mica motes where the radio transceiver uses the Chipcon CC2420 IEEE 802.15.4 (ZigBee) compliant chipset. This will allow the MicaZ motes to communicate with other ZigBee compliant equipment. The software stack includes a MicaZ mote-specific layer with ZigBee support and platform device drivers, as well as a network layer for topology establishment and single / multi-hop routing features. The MicaZ mote platform is built around the Atmel AtMega128L processor which is capable of running at 7.37 MHz. The MicaZ motes have 128 Kbytes of program memory, 512 Kbytes of flash data logger memory, and 4 Kbytes of SRAM. Power is provided by two AA batteries and the devices have a battery life of roughly 1 year depending on the application (very low duty cycle assumed). Sensor boards can be attached through a surface mount 51 pin connector, Inter-IC (I2C), Digital Input Output (DIO), Universal Asynchronous Receiver Transmitter (UART) and a multiplexed address/data bus.

Fig. 1. Crossbow MicaZ.

Here we present three projects in which Smartdust devices were used in the transport domain. In the first project, the UK Department for Transport funded ASTRA (ASTRA, 2005) (Applications of Smartdust in Transport) project, a preliminary investigation on Smartdust's feasibility was undertaken, with an aim to find out Smartdust's characteristics in the transport domain. The other two projects TRACKSS (TRACKSS, 2007) (Technologies for Road Advanced Cooperative Knowledge Sharing Sensors) and EMMA (EMMA, 2007)

(Embedded Middleware in Mobility Applications), both were EC FP6 funded) built on the information gathered from ASTRA project by implementing several ITS applications using Smartdust. The TRACKSS project focussed on achieving this through collaboration between Smartdust and other sensing technologies, whereas the EMMA project designed and implemented a middleware that enables integration of Smartdust with heterogeneous sensor platform in the transport domain.

3.1 ASTRA project

The ASTRA project investigated the use of mobile wireless sensor networks, and more specifically, Smartdust for transport applications. The project examined the current state-of-the-art with Smartdust devices, using MicaZ and Mica2 motes as the technology to be tested. It also looked at the likely market and technological advances of the Smartdust technology over the coming decade.

3.1.1 ASTRA project trials

A trial using Smartdust technology was hosted in Newcastle with a pervasive intelligent corridor established by a network of fixed motes on roads near Newcastle Central Station. Mobile motes were also placed in several buses. Communication between a static mote and a moving mote on-board a vehicle was achieved, showing that communication can take place between road side and vehicles using a network of motes. The architecture consisted of a number of Fixed motes (F) attached to bus stops along a circular bus route and a number of Mobile motes (M) carried by buses travelling in both directions along this route. One special fixed mote was placed in an interface board connected to a Data Server PC. The PC was connected to other Client PCs via the Internet. The range of the motes was such that it was unlikely that fixed nodes would always be able to communicate directly. They communicated with mobile nodes as buses passed bus stops. The buses communicated with each other as they passed. Together, the fixed and mobile nodes formed an ad-hoc network.

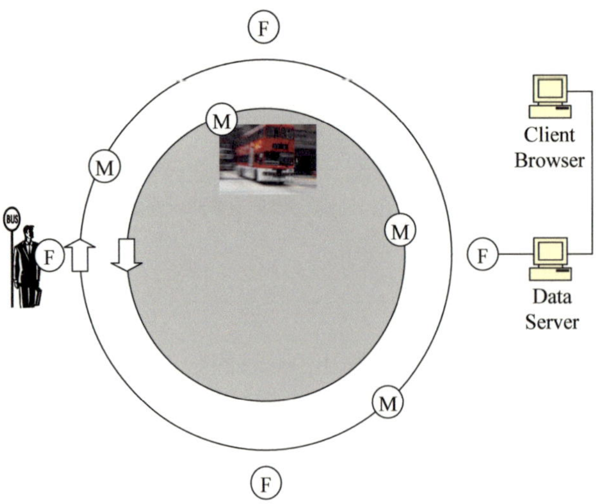

Fig. 2. ASTRA Experimental Setup.

The motes were concentrated in an area of Newcastle close to the Central Station, as shown in Figure 3 below. The red dots represent the bus stops which carried the motes. The buses travelled along Grainger Street and Neville Street before crossing the River Tyne to Gateshead (as seen in the map). The PC which acted as a Data Server was located in the Centre for Life, in the lower left-hand corner of Figure 3.

3.1.2 Experimental results

In an urban road environment radio wave reflections from surrounding buildings and objects can have a significant impact on the performance of the radio. In some situations the performance of the radio is enhanced and in other situations the performance is reduced. Reducing the motes radio transmitter power is likely to improve the performance of the radio in some cases due to a corresponding reduction in the number of radio wave reflections. Possible future research includes developing a way for motes to adjust their radio power to obtain optimal radio performance at a given location.

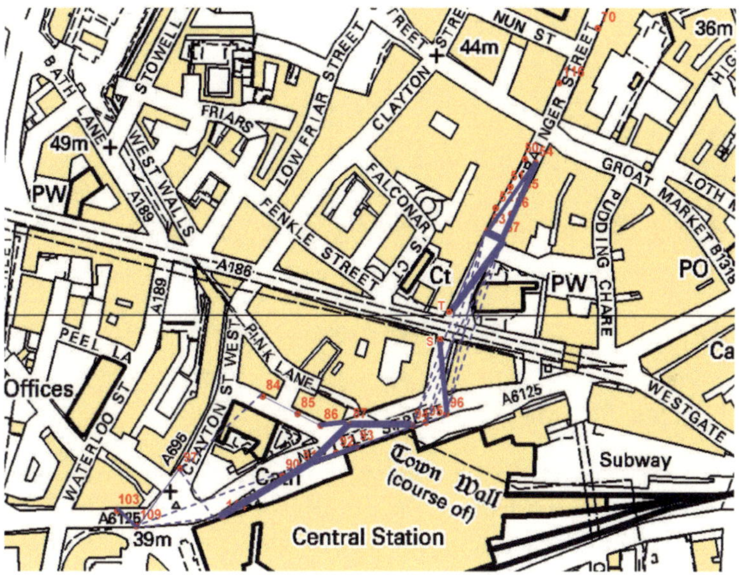

Fig. 3. Wireless sensor networks intelligent corridor.

The map is presented above is the Central Station area which shows the quality of each link. The quality of a link is based on the percentage of packets received, the higher the percentage the higher the link quality. This is indicated on the map by varying the width of the line between points where larger widths indicates higher quality links. A dashed line indicates total packet loss. The results also show that the distance between nodes is not the only factor in determining link quality. Radio reflections and interference seem to have played a significant role in determining the link quality during this experiment. Many of the shorter links did not show any connection whereas some longer links showed a good quality connection.

Experiments also showed that the MicaZ motes outperform the Mica2 motes in rural and urban environment due to their improved radio performance. Deploying MicaZ motes as opposed to Mica2 motes may therefore require fewer motes to cover the same geographical area. Matching the correct motes for the environment in which they are used will clearly be important. Although not tested within the trial, it is reasonable to conclude that the use of Wi-Fi or Bluetooth would result in a worse performance than using ZigBee as the former two technologies carry a greater overhead for new nodes joining a network.

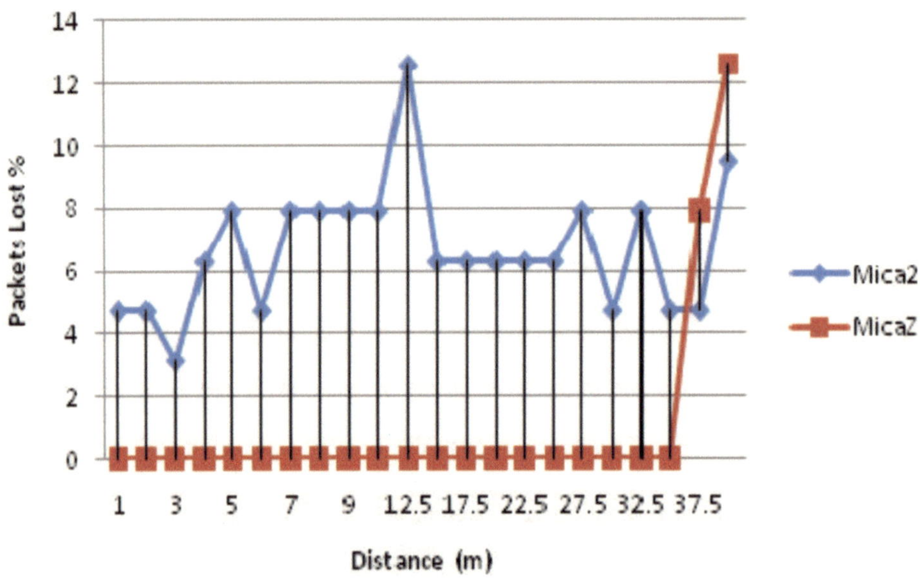

Fig. 4. Packets lost MicaZ vs Mica2.

Several experiments showed that it is possible to create connections between moving vehicles and a roadside infrastructure in an urban environment. It was also possible vehicles travelling at high speed (70mph). The time available for data transfer during a passage is a function of the radio range, the speed of the vehicle and the overheads involved in having the mobile node join the network. The ASTRA trial findings are summarized in Table 2.

Experiments were carried out to investigate the effect of mote packaging, height, antennae position and position inside the vehicle (Bus/Car) in radio range, received signal strength (RSSI) and packets loss. The experiment results showed these parameters have a significant effect on radio range and packets loss but can be overcome transmitting packets with high power level or using many devices to form a network (multi-hop technology). Experience gained from the ASTRA project trials provided the opportunity to use Smartdust devices in a range of transport prototype applications in the TRACKSS and the EMMA project. It seems that the opportunities of using Smartdust as part of a connected world (collaboration with other sensors/platform) for future intelligent transportation system are high.

	Range (m)	Max Power Signal (dBm)	% Lost Packets
Weather	Range is greater at lower temperatures (9.0°C - 100m, 17.1°C - 85m).	No difference.	No difference except towards the limits of range where packet loss rates increase as expected.
Packaging	The thicker the material of the packaging the shorter the range (waterproof box: 65m vs. trunking: 100m).	Thicker materials absorb more radio waves resulting in lower signal strength (The thicker waterproof box displays an average signal strength reduction of 15 dBm when compared with the trunking).	No difference except towards the limits of range where packet loss rates increase as expected.
Urban environment	Interference to the radio (caused by buildings, cars, people, etc.) has an effect on range.	Interference to the radio (caused by buildings, cars, people, etc.) has an effect on signal strength.	Interference to the radio (caused by buildings, cars, people, etc.) has an effect on packet loss.
Position (height)	The height of the nodes has an effect on range. The higher the node the greater the range (At a height of 0m the range is between 5m and 10m, at heights of 1m and 2m the range is at least 25m).	The height of the nodes has an effect on signal strength. The higher the node the higher the signal strength (at a distance of 25m the signal strength at heights 0m/1m/2m was NONE/-98.5/-94.1)	The height of the nodes has an effect on packet loss. The higher the node the lower the number of packets lost (at a distance of 25m the packet loss at heights 0m/1m/2m was 100%/37.7%/11.4%)
Position (antennae)		The orientation of the antennae has an effect on the received signal strength. The best orientation of the antenna is north and the worst orientation is south (North: -63.1 dBm, South: -80.4 dBm)	The orientation of the antennae does have an effect on packet loss. The best orientation of the antenna is north and the worst orientation is east (North: 0%, East: -2.4%)
Effect of chosen radio frequency		Choice of radio frequency has an effect on the received signal strength. In general lower frequencies result in a higher received signal strength over the same distance (868MHz: -57.9 dBm, 902MHz: -66.4 dBm, 928MHz: -72.7 dBm).	Choice of frequency in general does not affect packet loss. However interference from other devices using the same frequency can affect packet loss.

Table 2. ASTRA trial summary.

3.2 TRACKSS project

The focus of the TRACKSS (Technologies for Road Advanced Cooperative Knowledge Sharing Sensors) project was to research advanced communications concepts, open interoperable and scalable system architectures that allow easy upgrading, advanced sensor infrastructure, dependable software, robust positioning technologies and their integration into intelligent co-operative systems to support a range of core functions in the areas of road and vehicle safety and traffic management and control (TRACKSS, 2007). The overall aim was to develop new systems for cooperative sensing and predict flows, infrastructure and environmental conditions surrounding traffic, with a view to improving road transport safety and efficiency. Figure 5 shows how TRACKSS can be used in improving collaborations among various ITS sensors.

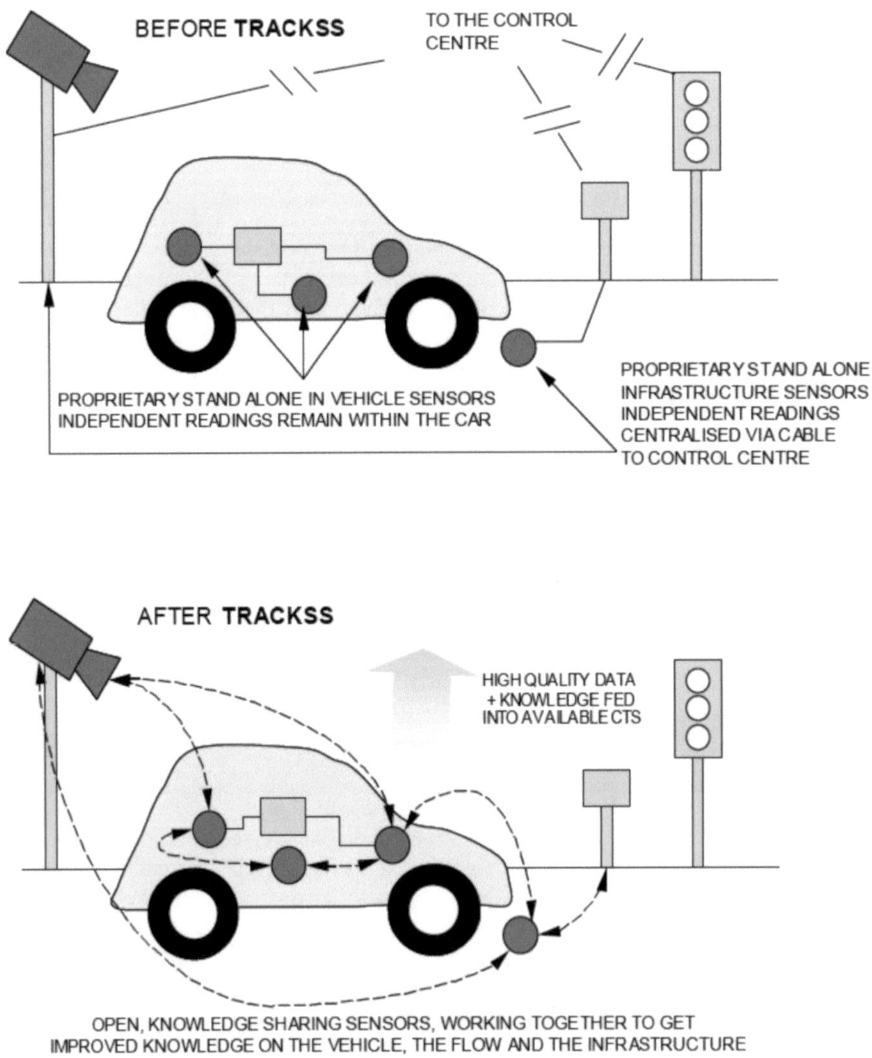

Fig. 5. How TRACKSS enabled sensor collaborations.

One of the major results of the TRACKSS project is the framework that provides support that allows communication between various sensors to be implemented readily and in a straightforward manner. This framework was designed to enable collaborations among infrastructure sensors, vehicle sensors, as well as vehicle and infrastructure sensors. Multiple scenarios have been designed, describing how the various sensors might collaborate through infrastructure, vehicle and mixed applications. There are two important concepts defined and implemented in TRACKSS:

Knowledge Sharing Model (KSM)

The KSM serves as the core of the framework, defining the XML format of the data to be exchanged among sensors, and providing a common API for communication. Data are classified into several pre-defined types of knowledge, and they are passed around using "publish and subscribe" mechanism supported by the KSM. The KSM platform is implemented through two executables: *ServerSAM* (which acts as the coordinating server, through which all communication is processed) and *TrackssRouter* (which handles the communication related to each sensor).

Knowledge Sharing Sensor (KSS)

In TRACKSS, each sensor is turned into a KSS by integrating the common communication protocols into its corresponding software. Each KSS can publish and/or subscribe to certain types of knowledge; the communication details are handled automatically by the KSM platform. Figure 6 presents the simplified version of the communication mechanism between two KSSs in TRACKSS. More details on TRACKSS KSM and KSS can be found in [24].

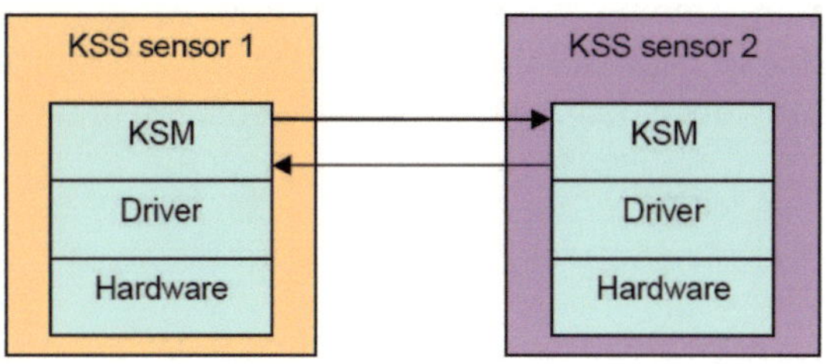

Fig. 6. TRACKSS KSM and its support for KSS.

In this paper, one of the TRACKSS collaborations involving Smartdust and Near-Infrared Optical Identification sensor is presented.

The Near-Infrared Optical Identification sensor was developed by LIVIC-LCPC (France) and its purpose is to detect, identify objects (i.e. vehicles or roadside objects) and localize them in the road scene. It is composed of two parts: an Emitter part and a Receiver part. The Emitter, which is the active part, is a near-infrared LED-based lamp, which, thanks to an embedded controller, codes an identifier (a number) using a defined frame protocol. The signal is thus time-coded (blinking light) and not space-coded. This enables important improvements with regard to detection ranges and robustness when compared to traditional spatial-pattern-based identification systems. The Receiver is a high-frame-rate, low-resolution CCD camera, equipped with an IR-bandpass filter, plus a decoding algorithm. By attaching an Emitter (with its unique identifier) to an object (such as a road sign), the Receiver can detect, identify, and locate it on the road scene. Full details of the Near-Infrared Optical Identification sensor can be found in (von Arnim, 2007 and 2008)

This collaboration was implemented using the support provided by the KSM and its executable platform, turning the Smartdust and the Near-Infrared Optical Identification sensors into two collaborating KSSs. Among others, this collaboration enables a more robust *Road Sign Recognition Application* to be implemented, where one sensor's capabilities can complement the other's weaknesses. This is based on the fact that no individual sensor is perfect: each has its own features and limitations. In this case, the Near-Infrared Optical Identification sensor has a very good range and localization features, but it relies on visual detection. This makes it prone to errors when objects may obstruct the view, or in inclement weather conditions. On the other hand, the Smartdust sensor cannot easily determine the location of a target (since it relies on radio signal, which does not necessarily carry the information on the detection direction), but the radio signal is free from visual impairment. By combining the positive characteristics of these two sensors, we have developed a new system that provides a more robust detection and identification of objects on the road.

3.2.1 Cooperative road sign detection application

By attaching the Smartdust transmitter and the Near-Infrared Optical Identification sensor Emitter to a road sign, suitably equipped vehicles can detect, identify, and localize the road sign ahead with sufficient time and even in poor visibility. For this application, we equip a car with a Near-Infrared Optical Identification Receiver (camera), a Smartdust device (sitting on its base station), and a computer connected to these sensors, on which the applications controlling these sensors are running. We also place infrared Emitters and Smartdust devices next to the sign on the side of the road; they act as the broadcaster of signal for representing the road signs. Figure 7 provides a diagrammatical representation of this set-up.

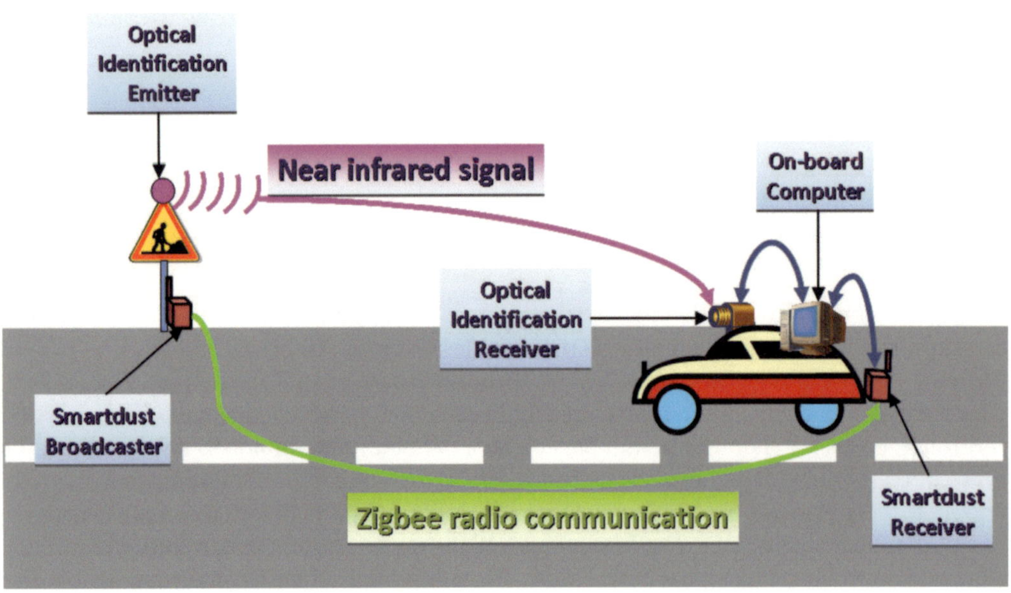

Fig. 7. The set-up of the I2V equipments.

On the on-board computer, both sensors communicate using TRACKSS' KSM. Essentially, the KSM serves a middleware that encapsulates a common communication protocol of these two sensors, enabling them to exchange information in a straightforward and uniform manner. This collaboration combines the long detection-range capabilities of the Near-Infrared camera with the visual-independence feature of the Smartdust in order to develop a more robust system for road sign detection.

In most cases, the Near-Infrared camera picks up the signal from the infrared Emitter prior to the Smartdust base station picking the signal from the Smartdust device on the side of the road. In these cases, the latter provides a confirmation of the detected road sign to the former, hence increasing the confidence of the result. In other cases (for example, around the bend or if there is some visual impairment), the Smartdust picks up the road sign signal before the camera. The system will then warn the driver that a road sign is detected nearby, but we do not have 100% certainty until this is confirmed by the camera.

Two examples of graphical displays can be seen on Figure 8. The left one shows a symbolic representation of the road sign and the right one show an "augmented reality" view, where the road sign is drawn at its real position on the image. One can easily understand the benefit of this application for night-time driver assistance. Further details of this collaboration - along with details of several other collaborations - can be found in (Arief & von Arnim, 2008, von Arnim et al, 2008).

Fig. 8. Graphical display for road sign recognition.

3.2.2 Experimental results

The first major experimental result of this application is given by the fusion of identification information from two different sensors with different technologies. When the fusion strategy is a confirmation of one sensor by the other, then the limitations of each sensor can be compensated.

Concerning the detection range, with our confirmation strategy (where the target must be within range to be simultaneously detected by both sensors), the final range is highly influenced by the range of either sensor, depending on weather conditions. The Near-Infrared Optical Identification sensor has a much higher range than the Smartdust sensor in its best-case scenario (low-luminosity). On the contrary, in the worst-case scenario for Near-Infrared Optical Identification sensor (sunny weather), the Smartdust usually has greater range. Another interesting observation is that the Near-Infrared Optical Identification sensor performs better when the vehicle is travelling at a high speed whereas Smartdust is better at low speed. So to sum up, both sensors complement each other in their capabilities. Our applications do not need a range over hundreds of meters. The fused sensors need to be able to detect a road sign sufficiently early to warn the driver or start an action, at any speed up to 130km/h. The ranges measured in both scenarios are sufficient. It is also important to note that the range of the Smartdust sensor can be easily increased by chaining multiple Smartdust sensors in an ad-hoc network ahead of the road sign. It is therefore possible for both sensors to have a comparable detection range.

From these detection range data, we can also derive information about the "distance-to-be-covered and available-time before passing the road sign". This value is very much affected by the speed of the vehicle; Table 3 shows the values at two different speeds: 50 km/h and 130 km/h. The road sign ID relates to the identification number of the road signs – in this case: 7 (representing a danger sign), 9 (representing an end of 50 km/h speed limit) and 20 (representing a traffic light).

Speed	Road sign	Distance (m)	Available time
50	7	141.7	10.2
50	9	133.3	9.6
50	20	125.0	9.0
130	7	89.5	2.4
130	9	135.2	3.7
130	20	95.5	2.6

Table 3. Distance and Time Range of Road Sign Detection.

These values compare favourably to the standard driver's reaction time of 1 second. Even at 130 km/h, we have at least 2.4 seconds anticipation time before the road sign.

3.3 EMMA project

The EMMA (Embedded Middleware in Mobility Applications) project (EMMA, 2007) had an overarching goal of utilising new embedded middleware to support the underlying logic and communications required for future cooperating wireless objects and the applications they may support in the automotive and road transport domains. The EMMA project was committed to deliver a middleware platform and a development environment which facilitates the design and implementation of embedded software for cooperative sensing objects. The ultimate aim of the project was to hide the complexity of the underlying infrastructure whilst providing open interfaces to 3rd parties enabling the faster, cost-efficient development of new cooperative sensing applications.

Fig. 9. EMMA hierarchical approach.

The EMMA network architecture can be considered at three levels (Figure 9): Within an automotive subsystem, at a vehicle level and at the supra-vehicle level. Recently, many wireless micro sensor network applications are being developed for a variety of applications including transport monitoring and control. However, there are still numerous challenges to be overcome if wireless sensor devices are to communicate with each other in an intelligent, cost effective and reliable way. The EMMA project selected the principle of hierarchical Wireless Co-operating objects (WICOs). The number and type of objects present in any given WICO hierarchy may vary depending on the prevailing conditions and their availability; however the handling of this can be abstracted away from the application developer by the middleware wherever possible.

It was necessary to find suitable technologies to support and coordinate WICOs at all three levels. Given this wide variety of sensor technology available and in common use, a popular approach to manage the growing complexity of a typical development is to create a malleable interface often called termed middleware. Middleware has been around for many years in different forms. "Middleware" in the loosest terms covers a wide variety of system integration schemes.

Evaluation of many middleware technologies has often reached the conclusion that middleware is a large and complex system, requiring careful coupling to, and tuning for, the specified application domain (EMMA, 2007). As a result there is a need to identify and develop a middleware system optimised for the challenges present in the intelligent transportation domain. A differentiating factor between current middleware models and the proposed EMMA system is the need to operate on multiple different platforms, with substantially different resources available.

Current middleware technology is typified by systems designed to run in non-real time, resource rich environments, such as large business information systems. A typical traffic infrastructure or automotive application runs in a real-time environment with strictly limited resources, distributed across many control units, themselves incorporating multiple dissimilar sensors (Hadim and Mohamed, 2006, Katramados et al, 2008).

EMMA middleware is capable of integrating systems for transportation application using large powerful fully featured PowerPC based FPGA technology (Xilinx ML403) which uses Qplus operating systems, as well as compact, battery-powered Smartdust devices which uses NanoQplus operating systems. An additional capability of the EMMA middleware is the ability to handle dynamic, ad-hoc networks. These are a natural result of communications between moving vehicle-based environments and infrastructure systems with finite static coverage. It must be possible to handle the potential changes to the network composition whilst insulating the application developer from the technical challenges that this involves.

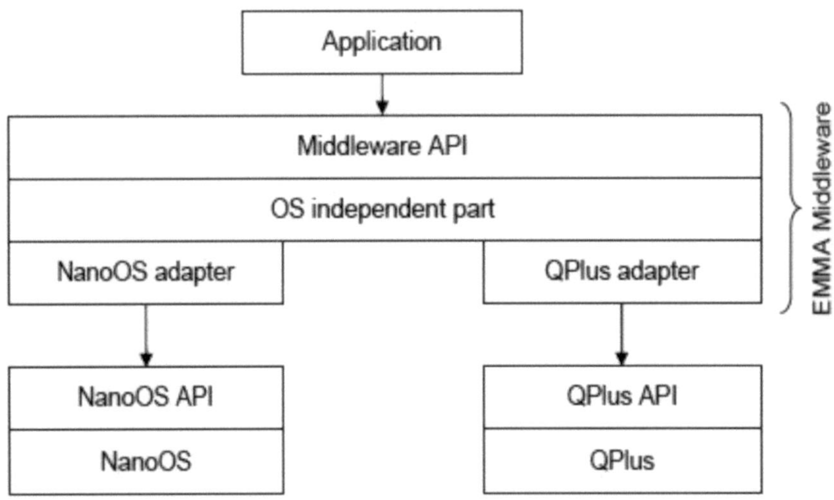

Fig. 10. EMMA Middleware.

The core part of the middleware is independent from the operating system and platform it is running on. This part provides several functions to the application through the middleware API. Since the different platforms vary strongly in resources like memory size, energy or CPU speed, not all functions will be provided on all platforms. The OS independent part, therefore, has to be modular. The adapters also include conversions for the APIs of different communication technologies, e.g. ZigBee or CAN. Each communication technology uses, for example, a different addressing scheme that has to be hidden from the EMMA applications. Therefore, the middleware defines an independent addressing.

The EMMA middleware is designed in a modular fashion. The communication adapters form the interface between the communication module and the actual hardware drivers. The communication module has a generic part and two specialised parts for message and data-centric communication (Yang & Wang, 2007). The detailed description of each of the modules is provided below.

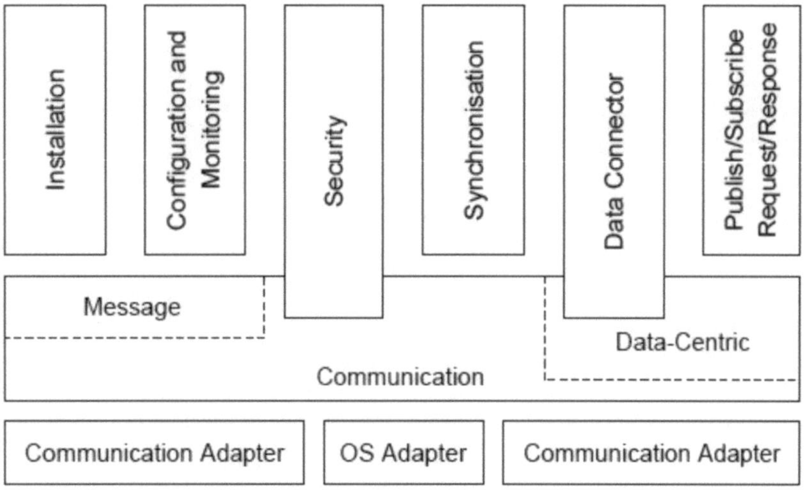

Fig. 11. Modular Components of EMMA Middleware.

Security is also an integral part of any ITS application which can be achieved with the combination of three core components, namely authentication, encryption and mechanisms for access control. The EMMA middleware provided both conceptual support for pre-distributed symmetric keys and conceptual support for symmetric encryption but no support for access control mechanism. Another FP7 funded PECES (PECES, 2009) project (a follow-up project of the EMMA project) is currently developing a complete security suite for pervasive computing applications including transport domain. The PECES project security suite make use of existing encryption protocols and standards to ensure the authenticity of devices, to exchange encryption key between communicating devices and to provide appropriate access control mechanisms to limit information sharing in and between devices. More detail information about PECES project security concepts can be found in PECES project Secure Middleware Specification deliverable (Handte et al, 2010).

The security module in the EMMA middleware provides management of keys and encryption and decryption function. The actual functions that perform encryption and decryption have to be provided by the application developers but the EMMA middleware provide mechanisms for calling them at the right time. Before a received message is delivered to the application, the security module is called to decrypt the message using the provided key and decryption function. Conversely, it is also called before handing over the message to the communication layer for sending.

In the road infrastructure domain, the scalability of the EMMA middleware allows it to be embedded in a variety of platforms from tiny Smartdust sensors to traffic control systems. By developing these devices using a common middleware based platform, the individual Smartdust devices can be configured as Wireless Cooperating Objects (WICOs). This WICO based sensor network allows the system to be flexibly configured to take advantage of changing conditions, for example with transitory sensors joining or leaving the fixed network as appropriate. The EMMA middleware provides a whole new range of opportunities for exchanging data amongst different vehicle sensors, subsystems and the engine. Numerous applications are continuously being developed aiming to improve the safety and enhance the driving experience. However, often such systems are isolated from the rest of the vehicle and their sensor information cannot be reused. As a solution, the EMMA middleware transforms such sensors to cooperating objects, giving the opportunity to develop new data fusion applications.

3.3.1 Vehicle to infrastructure application

Applications were developed in order to demonstrate the benefits of the middleware for giving priority to emergency vehicles. For the infrastructure level application, an emergency vehicle (ambulance, fire engine or police car, etc) are equipped with a Smartdust WICO (or a MicaZ mote installed with the EMMA middleware) which broadcast a beacon message if it is on an emergency situation. For example, in a busy intersection controlled by traffic lights, emergency vehicles are detected and given priority by regulating the state of the traffic lights (*Figure* 12).

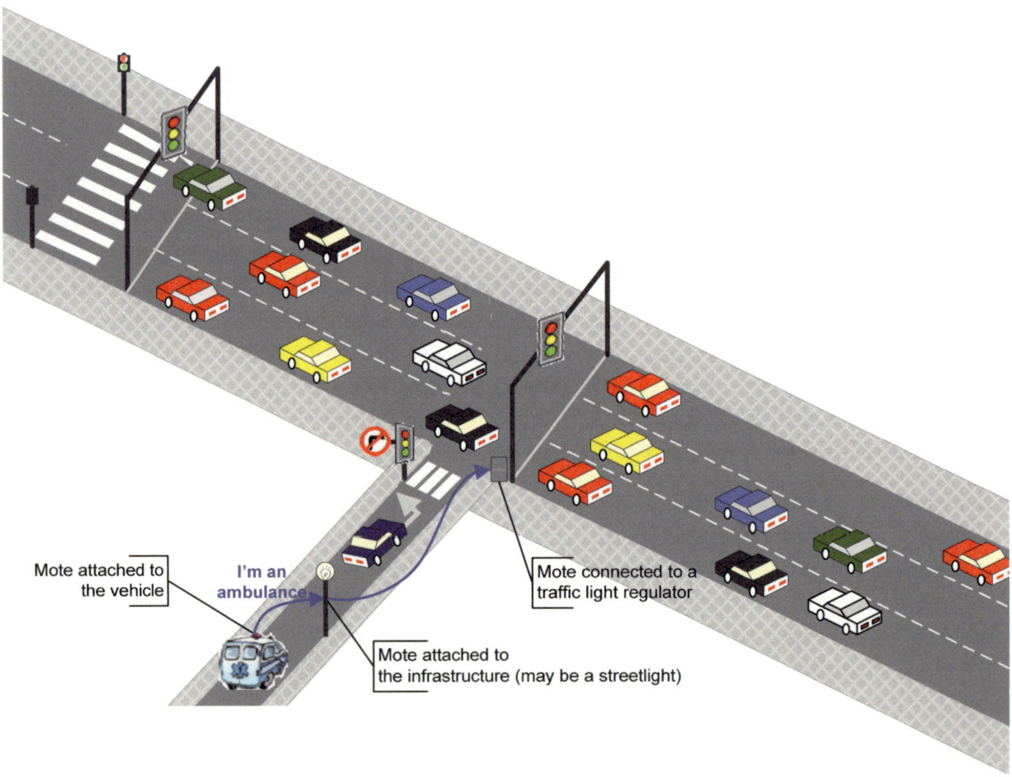

Fig. 12. Giving priorities for Emergency Vehicle.

In the Vehicle to Infrastructure application demonstration, a Smartdust WICO connected to a traffic regulator that acted directly by providing information to the regulator about an emergency situation. Smartdust WICOs were placed in the infrastructure to relay this message to the traffic light regulator. It was placed so far as needed in order to give time to the traffic regulator to change its status taking into account the time lost due to communication mechanisms (publish/subscribe) and other time periods the traffic regulator needs in order to guarantee safety first. The traffic regulator has been programmed to attend to the trigger signal provided by the mote activating an emergency control sequence. The demonstration has successfully carried out real road environment in Valencia, Spain.

An inter-hierarchical application, integrating all three levels that developed in the project: the automotive subsystem, the car level and the supra-car level, was also developed to demonstrate how heterogeneity issues can be solved in wireless sensor networks developing EMMA middleware. In this application demonstration, inter-hierarchical collaboration of the WICOs developed on the project consisted in transforming the information provided by the vehicles at both engine level (cylinder pressure and oil pressure sensors are based on TelosB platform) and vehicle level (GPS, Vehicle dynamic and Radar sensors are based on Xilinx ML 403 platform (Katramados et al, 2008)) into specific traffic control actions at infrastructure (to Smartdust WICO) level. In the infrastructure, a Smartdust WICO was connected to a portable VMS to display the information sent by the engine level and vehicle level sensors. The inter-hierarchical application scenario is explained in Figure 13 with the EMMA middleware components.

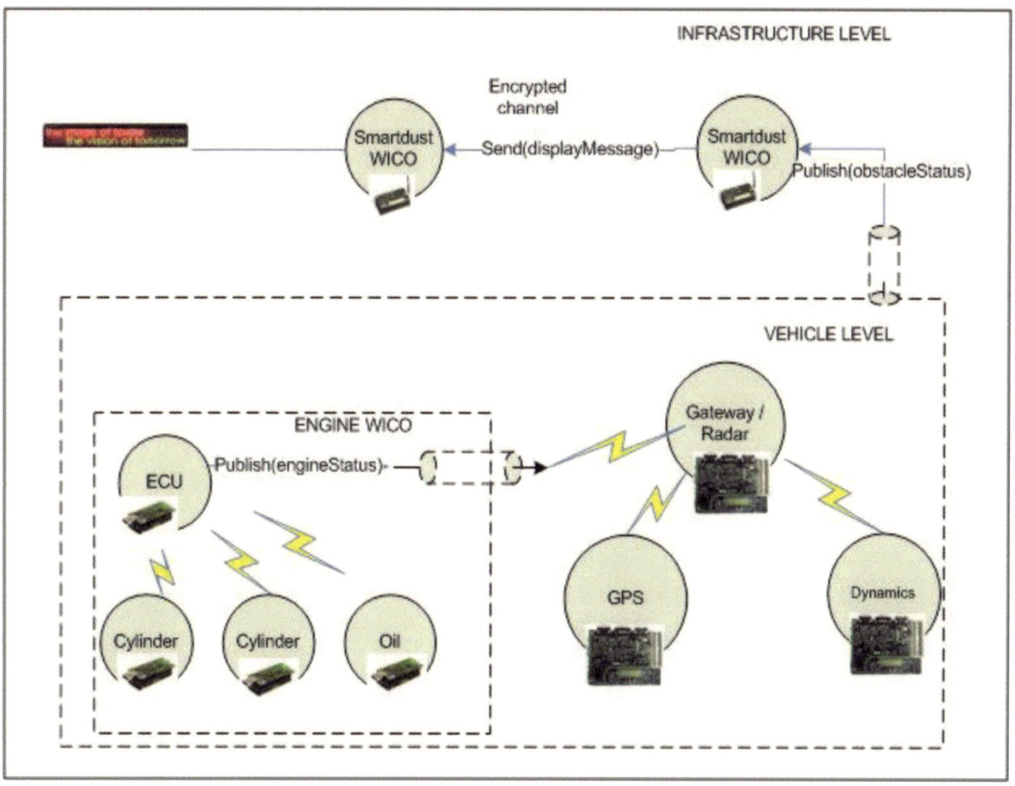

Fig. 13. Inter-hierarchical application.

The above application scenarios were implemented and tested with the EMMA middleware. This was used as a test bed to check the correct operation of EMMA approach for wireless sensor networks. It is worth mentioning that wireless pervasive computing applications for

ITS can be developed using the EMMA middleware but validation of such scenarios would require an extensive use of cooperative transport systems which is out of scope of the EMMA project. The EMMA middleware aims to provide an efficient platform to develop embedded applications for future intelligent transportation systems. The big advantage is that the EMMA approach is completely open to any systems and easily adapt to different applications.

3.3.2 Experimental results

Several set of experiments have been carried out with the EMMA middleware to evaluate possible use of Smartdust WICO in the supra-vehicle level as well as engine level and vehicle level application. These experiments intend to evaluate the use of wireless sensors with the EMMA Middleware for the proposed application scenarios at three different hierarchical levels. The engine level and vehicle level sensors are static networks and experiments showed satisfactory results for the application. The following section provides some experiments of the vehicle to infrastructure scenario. Two Smartdust WICOs were used and the first WICO was programmed with the EMMA Send message component and this WICO transmitted a packet for every, 500ms and 1000ms. The second WICO was programmed with the EMMA receive message component and connected to the MIB520 programming board which in turn was connected to a Laptop computer.

Open space experiment

This experiment was carried-out on the Exhibition Park, a large grassy area to the north of the Newcastle University. Each scenario, 100 packets were sent and those packets were received. Both WICOs were placed at 1m above the ground. The WICOs power level was set to maximum level. Each scenario repeated three times and calculations were performed offline to determine how many messages were lost at each distance and average values reported in Figure 14.

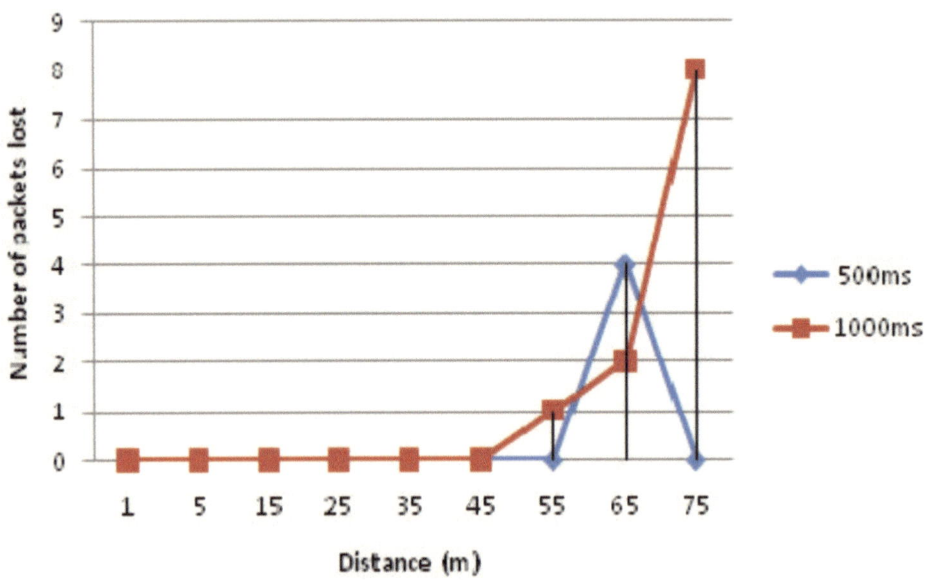

Fig. 14. Open space packets lost.

Urban environment experiment

This experiment was carried-out on the Claremont Road, a busy road near to the Newcastle University. Each scenario, 100 packets were sent and those packets were received. Both WICOs were placed at 1m above the ground. The WICOs power level was set to maximum level. Each scenario repeated three times and calculations were performed offline to determine how many messages were lost at each distance and average values reported in Figure 15.

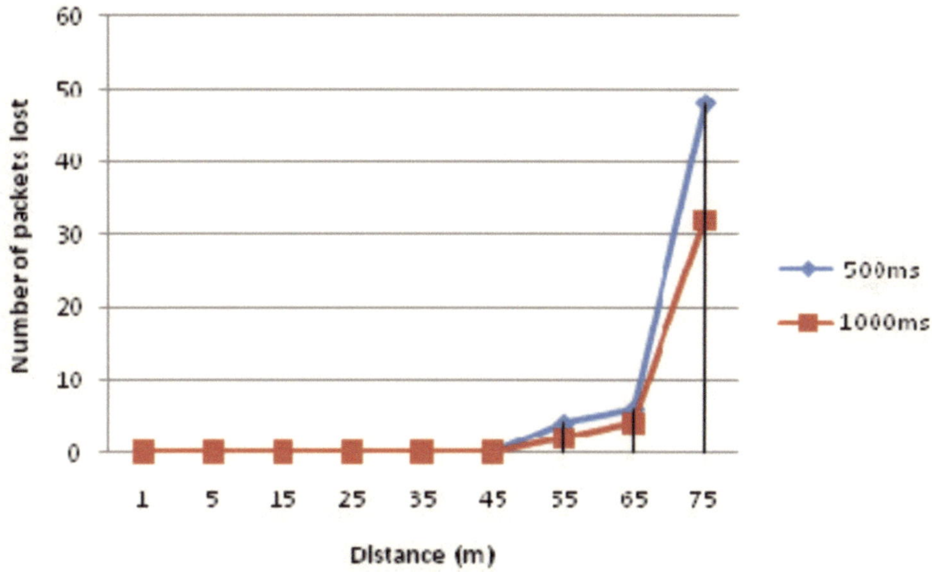

Fig. 15. Urban environment packets lost.

Mobile environment experiment

This experiment was carried-out on the Claremont Road up to 40mph and at the Motorway near to the Newcastle airport for higher speeds. The first Smartdust WICO was placed on a road side stand 1m from the ground and the second Smartdust WICO was placed on the middle of the dashboard of a vehicle. The Smartdust WICO at the vehicle sent message periodically which was received by the Smartdust at the road side. Each scenario repeated three times and calculations were performed offline to determine how many messages were lost at each distance and average values reported in Figure 16.

The open space experiment and urban environment experiment show that packets can be received without any packet lost up to 45m distance. The percentage of packets lost increases above 45m distance in both cases. In the mobile environment experiment, the received packets decrease with the speed increases as the WICO is in range for a shorter period of time. This means that communication time window decreases with the vehicle speed. At the 70mph speed, The WICO in the roadside received 5 and 11 packets for sending intervals 1000ms, 500ms respectively. And interestingly, there was no packets lost were seen between the first packet and the last packet received. The mobile environment experiment demonstrated that Smartdust WICO can be used with the EMMA middleware communication methods between a fixed infrastructure WICO and also fast moving vehicle-based WICO applications. This is an important finding which proves that the ZigBee Smartdust WICO do not suffer from any Doppler effects at normal motorway (70mph) speeds. It shows that the EMMA middleware communication functionality works well even the vehicle travelling at 70mph.

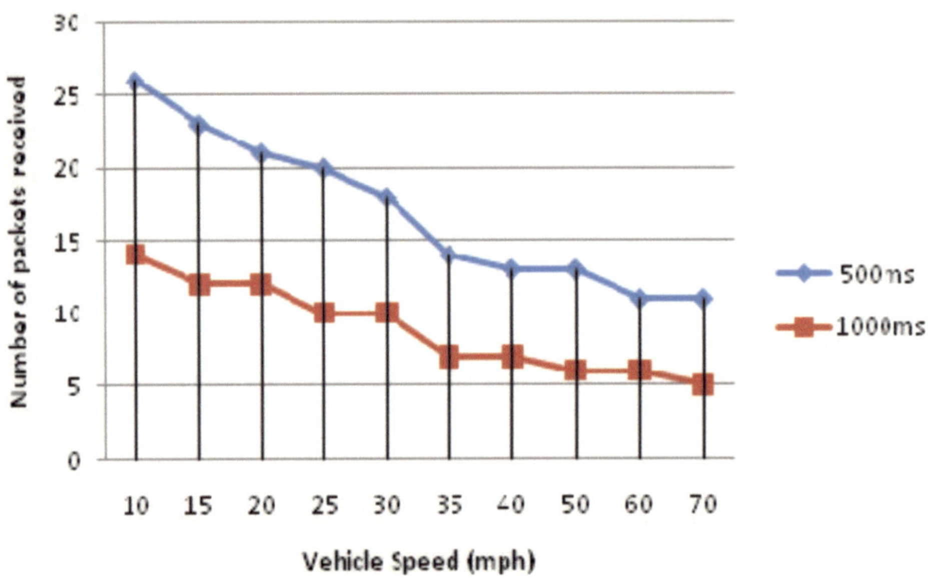

Fig. 16. Mobile environment packets lost.

4. Discussion

This chapter has summarised the research undertaken at Newcastle University to investigate the use of Smartdust devices for Intelligent Transportation System Applications. The costs of building and maintaining the infrastructure could be amortised over many such services delivered by Smartdust based wireless sensor networks. Research to deliver this concept of connected mobile devices and infrastructure leads to the opportunity to consider realistically for the first time a fully connected Intelligent Transport System for the future.

Research is currently focused on filling in the knowledge and technology gaps in pervasive, mobile ad-hoc wireless systems for a range of transport applications. Mobile wireless systems are beginning to be proven as a future tool that will enable the joining up of vehicles, individuals and infrastructure into a single 'connected' intelligent infrastructure system. Embedding this technology in infrastructure - such as environmental sensors in lampposts, embedded in vehicles and infrastructure, in goods, and even connecting individuals through their Smartphones, or even bespoke wearable wireless interfaces - offer potential for a more all-seeing, all knowing ITS infrastructure. If for example, vehicles are continually in wireless communication with the infrastructure, new paradigms for traffic monitoring and control could be considered, road space allocated more efficiently and incidents dealt with in an optimum way. If vulnerable users have such wireless devices, the infrastructure could warn vehicles to slow down and the drivers to be more vigilant -

indeed wireless devices attached to children could for example warn drivers that children are playing out on the street, just around the corner and to reduce speed now. Such devices could help with security and safety of individuals, be used on airline boarding cards and other tickets, and even be used to verify HOV (high occupancy vehicles) or blue badge entitlement. When such a system is also connected to say, a vehicle's CAN-bus, then information on driving style, strange driving behaviour (say where there is a badly maintained stretch of road or object in the road, could be detected from the CAN data - allowing mitigating and maintenance actions to be automatically triggered).

Many of these devices can carry payloads such as sensors, and the idea of monitoring pollution with these devices in a pervasive way is beginning to be researched by Newcastle University (with pervasive wireless environmental sensors being attached to lampposts). However if these devices become small and cheap enough (as is the future vision for Smartdust) then one could image that we each carry our personal exposure meter. Moreover with 'extreme' sensor design, wireless pollution sensors could be fitted in engine manifolds and exhaust pipes to allow the actual pollution generated by a vehicle to be measured and maybe adjustments to driving style or engine management systems can be advised or made to mitigate some of the pollution effects (early prototypes are being developed at the university at the moment). If future 'carbon allowances' are to be considered in the connected car, the pollution the car generates will also need to be measured and monitored - as proposed in the Smart Market Protocols project where auction and trading-based carbon allowances have been considered.

Significant research is required to fully realise the potential of such wireless systems, not just on the transport application side, but challenges to reduce the size of these devices from 'smart-lumps' to 'Smartdust' is critical as size, cost and power consumption of these devices will dictate whether the devices will become pervasive in the transport domain. This requires detailed work on antennae design, an investigation as to which is the most appropriate communications frequency, 802.11x, the influence of CALM, Wi-Fi and probably the most important challenge being battery power requirements (using power scavenging or other techniques).

The final key area of research which is still embryonic is in low-cost and robust sensor design - much work is on-going but uncoordinated in the transport domain. The robustness and dependability of mobile sensor devices, suitable communications protocols and e-science techniques to deal with the data are crucial. Also important is the issue of privacy and data protection in a potentially all-seeing, all-knowing connected world, which rise the question of how much information we want, need and what level of intrusion are we willing to bear.

5. Conclusion

It is clear that the next generation of vehicles will be required to have increased safety, lower emissions and more entertainment with higher performance than those of today. The innovations in wireless sensor devices and digital electronics will enable novel automotive applications which will become very common in future ITS. The challenges such as integrating heterogeneous devices for specific ITS application can be achieved by developing technologies such as in the TRACKSS and the EMMA projects.

This chapter has presented research projects that has been undertaken to investigate the suitability of using Smartdust and ZigBee technology for the intelligent systems applications. A selection of the applications and experiments carried out to systematically characterise the Smartdust technology in the road domain were presented here. The ability to communicate between vehicle and roadside illustrates that Smartdust will enable efficient and discrete communications between vehicle and roadside infrastructure - as the unit cost of Smartdust devices will continue to go down - this is a significant contribution to the ITS domain.

6. Acknowledgment

We would like to thank Axel von Arnim for his contribution to the Road Sign Recognition application during his time at LIVIC-LCPC. We also would like to thank Alberto Zambrano Galbis (ETRA R+D, Spain) and Carl Shooter (TRW Conekt, UK) for their contributions to the EMMA project vehicle to infrastructure level application. The work presented here was sponsored by various projects, including UK DfT-funded ASTRA project, EU-funded TRACKSS and EMMA projects.

7. References

Arief, B. & von Arnim, A. (2008) TRACKSS Approach to Improving Road Safety through Sensors Collaboration on Vehicle and in Infrastructure, *Proceedings of the 2nd IEEE International Symposium on Wireless Vehicular Communications (WiVeC 2008)*, Calgary, Canada, 21-22 Sep. 2008

Arief, B.; Blythe, P.T.; Fairchild, R.; Selvarajah, K. & Tully, A. (2007). Integrating Smartdust into Intelligent Transportation Systems, Technical Report CS-TR 1062, School of Computing Science, Newcastle University, Dec 2007

ASTRA(2005). "ASTRA Project" Project, last accessed 24 April 2011, Available from: http://research.cs.ncl.ac.uk/astra/

Bluetooth (2006). "Bluetooth (2006) Technology website", last accessed 24 April 2011, Available from: http://www.bluetooth.com/Bluetooth/Technology/

Blythe, P.T. (2006). Intelligent Infrastructure: A Smart Future with Smartdust and Smart Markets, *Proceedings of IFAC Conference on Transport automation and Control,* Delft, August 2006

Cai, H. & Lin, Y. A Roadside ITS Data Bus Prototype for Intelligent Highways, *IEEE Transaction on Intelligent Transportation Systems*, vol. 9, no. 2 , pp.344 - 348, June 2008.

Car2Car Consortium (2007). "Car2Car Consortium website", last accessed 30 April 2011, Available from: http://car-2-car.org/,

Chang, B. J; Huang, B. J. & Liang, Y. H. (2008). Wireless Sensor Network based adaptive vehicle navigation in multi- hop Relay WiMAX networks, *Proceeding of the 22nd International Conference on Advanced Information Networking and Applications*, March 2008.

Chen, Y.; Cheng, L.; Chen, C. & Ma, J. (2009). Wireless Sensor Network for Data Sensing in Intelligent Transportation System, *Proceeding of the 69th IEEE Vehicular Technology Conference*, 2009

Cook, J. A .; Kolmanovsky, I.V.; McNamara, D.; Nelson, E.C. & Prasad, K.V. (2007). Control, Computing and Communications: Technologies for the Twenty-First Century Model T, *Proceedings of the IEEE*, vol. 95, no. 2, pp. 334 - 355, Feb. 2007.

Crossbow (2006). "MPR/MIB User's Manual", last accessed 21 April 2011, Available from: http://www.memsic.com/products/wireless-sensor-networks/wireless-modules.html,

EMMA (2007). "EMMA Project", last accessed 24 April 2011, Available from: http://www.emmaproject.eu/

ETSI (2009). "*ETSI website*", last accessed 10 July 2011, Available from: http://www.etsi.org/WebSite/Technologies/IntelligentTransportSystems.aspx

Hadim, S. & Mohamed, N. (2006). Middleware: middleware challenges and approaches for wireless sensor networks, *IEEE Distributed Systems online*, volume 7, no. 3, March 2006.

Handte, M.; Haroon, M.; Apolinarski, W.; Rak, Z.; Zambrano, A.; Selvarajah, K. (2010). D4.1 Secure Middleware Specification, Available from: http://www.ict-peces.eu/

Heddebaut, M.; Deniau, V. & Adouane, K. (2004) . In-vehicle WLAN radio-frequency communication characterization, *IEEE Transactions on Intelligent Transportation Systems*, volume 5, issue 2, page(s):114 - 121, June 2004

ISO TC204 Working Group (2008). "CALM Concept", last accessed 20 May 2011, Available from: http://www.isotc204wg16.org/concept

Katramados, I.; Barlow, A.; Selvarajah, K.; Shooter, C.; Tully, A. & Blythe, P. T. (2008). Heterogeneous sensor integration for intelligent transport systems, Proceeding of the *IET Road Transport Information and Control - RTIC 2008 and ITS United Kingdom Members Conference*, Manchester, UK, 20-22 May 2008.

Kong, F. & Tan, J. (2008). A collaboration based hybrid vehicular sensor network architecture, *Proceeding of the International Conference on Information on Information and Automation*, 2008

Kosch, T. ; Kulp, I.; Bechler, M.; Strassberger, M.; Weyl, B. & Lasowski, R. (2009). Communication architecture for cooperative systems in Europe - [Automotive networking series], *IEEE Communications Magazine*, vol. 47, no 5, pp.116 - 125, May2009.

Lee, S.; Yoon, D. & Ghosh, A. (2008). Intelligent parking lot application using wireless sensor networks, *Proceeding International Symposium on Collaborative Technologies and Systems*, 2008

Marqués, A. (2007). Cooperative Sensors Making use of a Common Knowledge Sharing Model, *Proceeding of the ITS World Congress*, Beijing, China, October 2007

NanoQplus and Qplus (2008). "NanoQplus and Qplus website", last accessed 21 April 2011, Available: http://www.qplus.or.kr

Navet, N.; Song, Y.; Simonot-Lion, F. & Wilwert, C. (2005). Trends in Automotive Communication Systems", *IEEE Proceedings*, vol. 93, no.6, pp.1204 - 1223, June 2005

PECES (2009). "PECES Project", last accessed 3 October 2011, Available from: http://www.ict-peces.eu/

Selvarajah, K.; Tully, A. & Blythe, P.T. (2008). Integrating Smartdust into the embedded middleware in mobility application (EMMA) project, *Proceedings of the IET Road Transport Information and Control - RTIC 2008 and ITS United Kingdom Members Conference*, Manchester, UK, 20-22 May 2008.

TinyOS (2006). "TinyOS website" last accessed 21 April 2011, Available from: http://www.tinyos.net/

TRACKSS (2007). "TRACKSS Project" last accessed 24 April 2011, Available from : http://www.trackss.net/

Tully, A. (2006). Pervasive tagging, sensors and data collection: a science and technology review for the foresight project on intelligent infrastructure systems, *IET Intelligent Transport Systems*, vol. 153, pp.129 - 146.

UK DfT(2007). "Intelligent Transport Systems (ITS) - The policy framework for the road sector", last accessed on 20 May 2011, Available from: http://webarchive.nationalarchives.gov.uk/+/http:/www.dft.gov.uk/about/strategy/transportstrategy/eddingtonstudy/

US DOT (2010). "IntelliDrive whitepaper", last accessed on 20 June 2011 . Available from: http://www.its.dot.gov/press/2010/vii2intellidrive.htm

US DOT (2011). "US Connected Vehicle Research program", last accessed on 20 June 2011, Available from: http://mobilesynergetics.com/

von Arnim, A. ; Perrollaz, M. & Chevreau, J-M. (2007) Système Optique d'Identification d'un Véhicule", patent #fr-0752663, 15 Jan 2007

von Arnim, A.; Arief, B. & Fusée, A. (2008). Cooperative Road Sign and Traffic Light Using Near Infrared Identification and ZigBee Smartdust Technologies, *Proceedings of the 15th World Congress on Intelligent Transport Systems (ITS 2008)*, New York, USA, 16-20 Nov 2008

von Arnim, A.; Perrollaz, M.; Betrand, A. & Ehrlich, J. (2007). Vehicle Identification Using Near Infrared Vision and Applications to Cooperative Perception, *Proceedings of the IEEE Intelligent Vehicles*, Istanbul, 13 Jun 2007

Warneke, B.; Last, M. & Pister, S. J. (2001). Smartdust: communication with a cubic-millimeter computer", *IEEE computer*, vol. 34, no. 1, pp. 44 - 51, Jan 2001.

Yang, L. & Giannakis, G. B. (2004) Ultra-wideband communications: an idea whose time has come, *IEEE Signal Processing Magazine*, vol. 21, no. 6, pp.26 - 45, Nov. 2004

Yang, L. & Wang, F-Y. (2007). Driving into Intelligent Spaces with Pervasive Communications, *IEEE Intelligent Systems*, vol. 22, no. 1, pp:12 - 15, Jan.-Feb. 2007

ZigBee Alliance (2006). "ZigBee Alliance website", last accessed 24 April 2011, Available from: http://zigbee.org/,

7

Active Traffic Management as a Tool for Addressing Traffic Congestion

Virginia P. Sisiopiku
University of Alabama at Birmingham
USA

1. Introduction

Recurrent and non-recurrent congestion in urban areas continues to be a major concern due to its adverse impacts on delays, fuel consumption and pollution, driver frustration, and traffic safety. In the U.S., limited public funding for roadway expansion and improvement projects, coupled with continued growth in travel along congested urban freeway corridors, creates a pressing need for innovative congestion management approaches.

Congestion management is certainly not a new concept. Strategies targeting demand management in response to recurrent congestion have been utilized for years. Examples include ramp metering, high occupancy vehicle (HOV) lanes and value pricing options. Other strategies focus on operational management such as incident detection and management in response to nonrecurring congestion. Many of these strategies rely on Intelligent Transportation Systems (ITS) for surveillance, congestion monitoring, and information dessemination and aim at increasing operational efficiency of existing systems.

Active Traffic Management (ATM) is a new approach that utilizes many of these same principles but does so in a unified manner in order to maximize the efficiency of transportation facilities during all periods of the day and under both recurrent and non-recurrent congestion conditions. ATM typically relies on comprehensive automated systems to continuously monitor and adjust roadway management strategies as traffic conditions change over time. This approach stresses automation, which in turn, allows for dynamic deployment of strategies to quickly optimize performance and enhance throughput and safety. Through continuous system monitoring, dynamic response, and coordination of systems, ATM provides a holistic approach to transportation system management (Sisiopiku et al., 2009).

2. Advanced traffic management options

ATM is composed of a set of different strategies that can work synergistically (or on an individual basis) to achieve the common goal of congestion alleviation. Strategies considered under the ATM umbrella include: speed harmonization, high occupancy vehicle lanes, dynamic junction control, and temporary shoulder use. Details on the principles of each of these options are provided below.

2.1 Speed harmonization

Speed harmonization systems use changeable speed limit signs posted over each lane to regulate freeway speeds based on prevailing traffic conditions. Speed limits can be adjusted when freeway conditions are unsuitable for high speed operations, such as under adverse weather conditions or low visibility. Speed limits can also be lowered when there is an incident or congestion on specific segments in order to reduce the chances of secondary accidents and facilitate a smoother flow of traffic.

Advanced versions of the speed harmonization strategy include dynamic implementation (based on real-time travel demand, not simply time of day) along with dynamic speed controls to improve the overall safety and efficiency of freeway operations. Through speed harmonization, agencies can make the most of existing capacity by delaying the point at which flow breaks down and stop-and-go conditions occur.

2.2 High Occupancy Vehicle (HOV) lanes

HOV lanes are lanes available to vehicles that meet a minimum occupancy requirement. The main purpose of HOV facilities is to maximize the passenger-carrying capacity of the roadway, especially in peak periods. Entrance restrictions typically apply to passenger vehicles carrying less than two persons. The use of HOV lanes by transit buses, vanpools, and carpools is encouraged to further increase the carrying capacity of HOV lanes and lighten the traffic load of adjacent general use lanes.

The main objective of HOV use is to reduce congestion and encourage people to carpool or vanpool. This behavior reduces air pollution and saves money (NCDOT, 2008). HOV lanes can be open 24 hours a day, 7 days a week, or managed dynamically, in which case they become part of ATM.

Often HOV lanes are utilized as High Occupancy Toll Lanes (HOT), allowing single-occupant vehicles to use HOV lanes during peak hours in return for a toll. Under the value priced management option, the tolls can change dynamically according to real-time traffic conditions and manage the maximum number of cars in the HOT lanes in order to keep the traffic lanes free of congestion, even during rush hour. Examples of states with proven applications of HOT lanes include California, Colorado, Florida, Minnesota, Texas, Utah, and Washington.

2.3 Junction control

The junction control strategy is a combination of ramp metering and lane control at on-ramps (Berman et al., 2006). Typically, junction control is applied at entrance ramps or at merge points where the number of downstream lanes is fewer than that of upstream lanes. In the U.S., this strategy has been applied statically by dropping one lane from the outside lanes or merging the two inside lanes (Tignor et al., 1999). In Germany and other European countries this is done dynamically by installing lane control signals over both upstream approaches before the merge. This strategy gives priority to the facility with the higher volume and results in a lane drop on the approach with the lower volume (Mirshahiet et al., 2007).

The objective of junction control through either static or real-time means is the better management of recurrent congestion by making traffic flow more uniform, utilizing more effectively the existing roadway capacity, and improving traffic safety. The literature reports decreases in primary collisions by 15 to 25 percent through implementation of junction control strategies (Stone et al., 2007).

2.4 Temporary shoulder lane use

Temporary use of shoulder lanes as travel lanes began in many cities in the late 1960s. This strategy is usually employed during peak periods and in the peak direction and gives permission to vehicles to use either the right or left shoulder lanes in specific conditions. Temporary shoulder lane provides additional lane(s) within the existing pavement, without the need to widen the freeway (FHWA, 2003). Under the ATM concept, the use of shoulder lanes is done by using dynamically variable signs to let drivers know that the shoulder lane is open in a certain segment. The purpose of temporary shoulder use is to improve the performance of freeway facility by providing additional capacity when needed.

Temporary use of shoulder lanes on freeways is a strategy currently employed in select U.S. cities to provide a temporary capacity increase for congested freeways during the times when demand is greatest. In a typical application, motorists are allowed to use shoulders as an extra driving lane during the AM and PM peak periods while in other deployments, freeway shoulder lanes are used by transit buses during certain periods of the day.

Temporary use of shoulder lanes is also employed as an active congestion management strategy in Europe. In several countries, it is coupled with speed harmonization to enhance its effectiveness. Speed harmonization systems allow freeway operators to reduce freeway speeds during times of shoulder lane usage to ensure that improved traffic operations are also coupled with reduction in the chance of occurrence and severity of crashes.

The use of the shoulder lanes requires the presence of traffic control devices in order to inform the users whether the shoulder lane is open or not. A variety of traffic control devices and other pertinent technologies should be utilized to ensure driver safety when opening the shoulder lane. These include lane control signals, dynamic speed limit signals, dynamic message signs, closed-circuit television cameras, roadway sensors, and emergency roadside telephones (Mirshahiet et al., 2007). An example of a system in operation utilizing several of the above mentioned technologies is depicted in Figure 1.

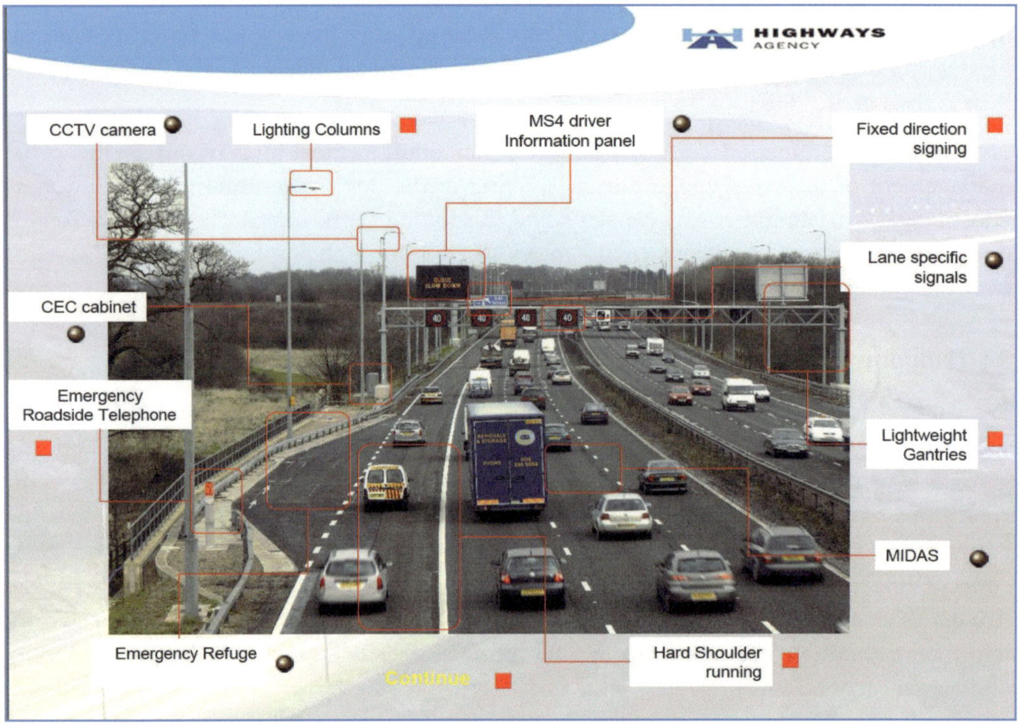

Fig. 1. Traffic Devices and Strategies in M24 UK (Stone et al., 2007).

2.5 Expected ATM benefits

European studies from Denmark, England, Germany, and the Netherlands confirm that ATM strategies result in great benefits, including increase in vehicle throughput, crash reduction, improvement in trip reliability, decrease in congestion and traffic delays, and an overall improvement in the driving experience. Depending on the location and the combination of strategies deployed, specific benefits measured in Europe as a result of this congestion management approach include the following (Mirshahiet et al., 2007):

- Increase in average throughput for congested periods of 3 to 7%
- Increase in overall capacity of 3% to 22%.
- Decrease in primary incidents of 3% to 30%.
- Decrease in secondary incidents of 40% to 50%.
- Overall harmonization of speeds during congested periods.
- Decreased headways and more uniform driver behavior.
- Increase in trip reliability, and
- Ability to delay the onset of freeway breakdown.

Based on international experience, a summary of potential benefits from deployment of a variety of ATM options in the U.S. is provided in Table 1.

Active Traffic Management Strategy	Potential Benefits												
	Increased throughput	Increased capacity	Decrease in primary incidents	Decrease in secondary incidents	Decrease in incident severity	More uniform speeds	Decreased headways	More uniform driver behavior	Increased trip reliability	Delay onset of freeway breakdown	Traffic noise reduction	Emissions reduction	Fuel consumption reduction
Speed harmonization	X		X		X	X	X	X	X	X	X	X	X
Temporary shoulder use	X	X							X	X			
Queue warning			X	X	X	X	X	X	X		X	X	X
Dynamic merge control	X	X	X			X		X	X	X	X	X	X
Construction site management	X	X							X		X	X	X
Dynamic truck restrictions	X	X				X		X	X			X	X
Dynamic rerouting & traveler info	X		X	X				X	X			X	X
Dynamic lane markings	X	X							X				
Automated speed enforcement			X		X	X		X	X			X	X

Table 1. Potential Benefits from ATM Implementation (Mirshahiet et al., 2007).

3. Advanced traffic management in the United States

3.1 Current state of practice

ATM received attention in the U.S. only in the recent years as an approach with great potential to address the ever-growing congestion problems in urban areas. A milestone in the development of ATM in the U.S. was the 2006 International Technology Scanning Program sponsored by the FHWA, the American Association of State Highway and Transportation Officials (AASHTO), and the National Cooperative Highway Research Program (NCHRP). Through this effort, a team of U.S. experts studied European ATM systems and concluded that ATM is the next evolution in congestion management in the U.S. In their report to FHWA, the experts identified nine key recommendations related to congestion management with a potential to ease congestion in the U.S., as follows (Mirshahiet et al., 2007):

- Promote ATM to optimize existing infrastructure during recurrent and non-recurrent congestion.
- Emphasize customer orientation and focus on trip reliability.
- Integrate active management into infrastructure planning and programming processes.
- Make operations a priority in planning, programming, and funding processes.
- Develop tools to support active management investment decisions.
- Consider public-private partnerships and innovative financing and delivery strategies.
- Provide consistent messages to roadway users.
- Consider pricing as only one component of a total management package, and
- Include managed lanes as part of the overall management of congested facilities.

A 2009 exploratory study by Sisiopiku et al. surveyed system managers from several state transportation agencies to identify current and planned ATM initiatives in the U.S. The study focused on four state agencies that had either implemented or were in the process of implementing ATM projects in their highways systems, namely Washington State DOT (WSDOT), Minnesota DOT (MnDOT), Virginia DOT (VDOT), and California DOT (CalTrans). A summary of ATM initiatives in each of these states as reported in Sisiopiku et al. (2009b) and updated to reflect recent developments follows.

3.1.1 Washington State DOT (WSDOT)

The Washington State DOT has two ATM projects recently completed: the SR 520/I-90 ATM Project and the I-5 Variable Speed Safety Project. The SR 520/I-90 project installed variable speed signs and lane control signals on the SR 520 and I-90 bridges over Lake Washington in Seattle. The goal was to improve speed control and better cope with incidents during the reconstruction of the SR 520 floating bridge. The I-5 project installed variable speed limit and lane control signs on the northbound lanes of I-5 approaching Seattle. This is a speed harmonization system that automatically adjusts speed limits based on prevailing traffic conditions to optimize throughput and safety and is operational since 2010.

The WSDOT is treating these two projects as the beginning of a broader ATM program and the outcomes are closely monitored. At present, funding is seen as one of the major obstacles to the expansion of this program. It should be noted that these ATM projects had unique funding sources: the I-5 project was funded as part of a mitigation plan for the reconstruction of the downtown Seattle viaduct. As yet there is no specific funding for additional ATM efforts.

The costs for each program are averaging about $4.0 million per mile, with lane control sign structures located approximately every ½ mile. WSDOT staff did not identify any substantial implementation issues with either project, although because both are the first of their kind in the state they are being monitored very closely.

3.1.2 Minnesota DOT (MnDOT)

The Minnesota DOT has more than one major ATM projects in operation. The I-35W project took place in 2009 in an effort to convert/construct HOT lanes on a 14-mile segment of I-35W near Minneapolis. Part of this effort required the implementation of a shoulder lane use program on the last two mile segment into Minneapolis. The shoulder lanes are used only during peak hours and require a toll, which can be varied based on prevailing traffic conditions. Overhead lane control signings are placed approximately every half mile and speed advisories are conveyed via standard CMS message signs.

The MnDOT obtained legislative approval to implement a shoulder use program for I-35W. This was the first shoulder lane use program in the state for passenger vehicles, although MnDOT has already implemented extensive shoulder use programs for transit vehicles. The experience with transit shoulder use has been excellent and helped with public education about the project, since motorists were already familiar with shoulder use during peak periods. The program was funded under an FHWA urban partnership agreement and is monitored extensively in an effort to assess the feasibility of future similar projects.

Moreover, variable speed limits are available since 2010 on I-35W between Bloomington and Burnsville. They are part of a real-time system called Smart Lanes, which is expected to expand on I-94 between St. Paul and Minneapolis by fall 2011.

3.1.3 Virginia DOT (VDOT)

The Virginia DOT has several ATM projects either in operation or in the planning stages located in the Northern Virginia/Washington, D.C. area. They include the use of shoulder lanes, variable speed limits, and the construction of HOT lanes in the I-459 corridor.

VDOT has a shoulder lane use program in operation in the I-66 corridor. Historically the shoulder lanes have been used only during strictly defined AM and PM peak periods, regardless of traffic conditions, and only in the peak direction. Recently, however, VDOT has extended the use of shoulder lanes to 5:30 - 11:00 in the morning and 2:00 - 8:00 in the afternoon to handle growing traffic congestion. More importantly, VDOT has also begun to allow shoulder lane use during major incidents or when construction causes lanes to be closed. This has effectively made the shoulder use program an ATM strategy.

Furthermore, VDOT uses variable speed limits (VSL) as part of the Woodrow Wilson Bridge project on I95/495. The variable speed limit signs have been used primarily to reduce vehicle speeds, improve traffic flow, and improve safety during periods of construction. The system is active in a 7 mile long segment of the I-95/495 corridor just west of the Wilson Bridge. The system has been successful enough that in May of 2009 VDOT decided to extend the use of the VSL system to the AM and PM peak periods, when it functions as a speed harmonization system. VDOT is currently collecting data on its effectiveness.

Another ATM initiative is the construction of HOT lanes on I-459 and I-95. While traditional HOV-3 vehicles and motorcycles will be able to use the lanes for free, non-HOV vehicles will be allowed to pay to use the lanes. Lane usage fees will vary depending on demand and road conditions. Project construction has begun but operation is not anticipated until 2013.

Discussions with VDOT staff indicated a commitment to implement ATM strategies. The major obstacle to implementation is seen as funding. ATM measures require extensive infrastructure for surveillance and traffic control; implementation of ATM strategies can prove quite costly and funding is not available for a larger program at this point. Nonetheless, reactions to initial projects have been positive.

Implementation issues to this point have been seen as manageable. VDOT has monitored traffic operations in the I-66 corridor to ensure that the use of shoulder lanes has not reduced safety. To date their experience has been similar to that of other agencies that have implemented shoulder lane programs, which have found minimal impacts to safety. VDOT staff indicated that new ATM programs will be monitored extensively to ensure that they are producing cost effective results.

3.1.4 California DOT (CalTrans)

In 2009, CalTrans was contacted to assess the status of ATM in the state. CalTrans staff said that there were no ATM projects active within the state, although the first was in the planning stages for Alameda County and the Bay Area. The I-80 Integrated Corridor Mobility (ICM) Project aims at incorporating several ATM strategies, namely adaptive ramp metering, variable speed limits (speed harmonization), and adaptive lane controls. These strategies would be combined to regulate the flow of traffic in the I-80 corridor in order to maximize throughput, minimize incidents, and better handle incidents when they occur.

CalTrans staff said there were no long term ATM plans beyond the current I-80 project and that the state's current financial problems are a limiting factor in that regard. Staff did indicate, however, that they felt ATM would become more important in state planning in the future. One of the primary areas of focus for managing congestion in the future will be HOV/HOT lanes and those will likely require extensive shoulder use programs to be feasible. CalTrans staff indicated that ATM would have to be an integral part of such programs. Staff also indicated that the State has interest in testing speed harmonization in the future, although there are no firm plans in the works.

Overall, the interviews with state agencies indicated that there is interest in implementing ATM strategies in the US in the future but the initial deployments would be limited. The primary implementation issues identified were a. funding, b. legal, and c. public education (Sisiopiku et al., 2009).

All state representatives interviewed said that available funding was a limiting factor in their decisions to develop ATM programs. Only a few projects have been initiated and most of the ones that are have unique funding sources (e.g., they are mitigation measures tied to a larger highway project). The infrastructure required for ATM strategies can be extensive so funding is an ongoing concern.

Although all agencies interviewed were able to obtain the legal and legislative clearances needed, some ATM strategies, such as shoulder lane use, raise questions about public safety and need to be addressed at a local level.

Efforts to educate the public about the proper use of ATM options are required, because many of the strategies are new to this country. VDOT, for example, has developed a public education campaign for the variable speed limit system at the I-459 Wilson Bridge Project. The campaign includes advertisements and a website explaining the purpose and function of the system. Because so few ATM projects have been implemented, an assessment of the best public education strategies would prove useful.

In general, the states interviewed did not see major technical hurdles to implementing ATM strategies. The technology, for the most part, was viewed as having been proven either in isolated ITS applications or in European ATM applications. The most likely initial applications of ATM technologies include shoulder lane use, speed harmonization, dynamic HOT lanes, and adaptive lane controls.

3.2 The future of ATM in the U.S.

Planning for ATM is an important ingredient for success. Whether or not to implement ATM and its operational strategies is a policy decision that must be made at the appropriate governing level. To that end, policymakers should develop both short- and long-range plans that incorporate ATM into the framework of transportation alternatives. Furthermore, agencies should approach ATM proactively by including it in current and future plans for target corridors. They should assess what ATM capabilities already exist in those corridors and what components need to be added to facilitate active management, even if conditions do not currently warrant such operational strategies. This forward-thinking approach will ensure that the infrastructure is put into place during future projects so that ATM can be implemented when warranted by congestion levels and mobility needs. In some regions, legislative support may be necessary to make this operational approach possible (Mirshahiet et al., 2007).

The technologies required for implementation are currently available in the market; however, research and development may be required to refine existing systems, and careful selection of available technologies should take place to improve cost/effectiveness while accounting for local needs and special conditions.

Implementation, operation, maintenance, enforcement, and marketing are some of the policy decisions that govern a successful ATM system. For example, potential policy decisions that would need to be made in tandem with the planning and implementation of temporary shoulder lane use are being reviewed and documented. Special emphasis should be placed on the development of a set of procedures that will clearly describe how to open and close the shoulder to traffic operation, and assign roles and responsibilities. Moreover, considerations related to maintenance, compliance, and enforcement, and institutional issues (such as regulatory and legal issues, finance, organization and management issues, and human resources) must be studied carefully (FHWA, 1994).

Another key consideration is the availability of funding. Therefore, funding resources at the federal, state, or local level that can potentially support design, implementation, operation, maintenance, and marketing of the project should be identified and committed.

In addition, various institutional issues are essential to the successful implementation of ATM and include customer orientation; the priority of operations in planning, programming, and funding processes; cost-effective investment decisions; public-private partnerships; and a desire for consistency across borders. These issues need to be considered carefully prior to implementation in order to maximize the potential for success.

3.3 Promising ATM options for deployment in the U.S.

The FHWA International Scan Tour Report (Mirshahiet et al., 2007) offered specific recommendations for ATM implementation in the U.S. in response to congestion. A summary of these recommendations follows.

3.3.1 Speed harmonization

The United States should implement speed harmonization on freeways as a strategy to actively manage the network and delay the onset of congestion under normal operating conditions. The system should include the following elements (Mirshahiet et al., 2007):

- Sufficient sensor deployment for traffic and weather monitoring to support the strategy.
- Adequate installation of sign gantries to ensure that at least one speed limit sign is in sight at all times.
- Placement of speed limit signs over each travel lane.
- An expert system that deploys the strategy based on prevailing roadway conditions without requiring operator intervention. It is critical that this expert system be reliable and accurate to gain the trust and acceptance of the public.
- Connection to a traffic management center that serves as the focal point for the system.
- Passage of enabling legislation and related laws to allow for dynamic speed limits.
- Uniform signing related to speed harmonization and its components.
- Modeling tools to assess the impacts of speed harmonization on overall network operations.
- Closed-circuit television (CCTV) cameras to support the monitoring of the system.
- Dynamic message signs to provide traveler information and regulatory signs as appropriate, and
- Automated speed enforcement to deter violations.

3.3.2 Temporary shoulder lane use

Temporary shoulder lane use should be implemented, where appropriate, to temporarily increase capacity during peak travel periods. Specific elements of the operational strategy should include the following (Mirshahiet et al., 2007):

- Deployment in conjunction with speed harmonization.
- Passage of enabling legislation and related laws to allow the shoulder to be used as a travel lane.
- A policy for uniform application of the strategy through entrance and exit ramps and at interchanges.
- Adequate installation of sign gantries to provide operational information and to ensure that they are in sight at all times.
- Placement of lane control signals over each travel lane.
- Uniform signing and markings related to temporary shoulder use.
- CCTV cameras with sufficient coverage to verify the clearance of the shoulder before deployment.
- Provision of pullouts at regular intervals with automatic vehicle detection to provide refuge areas for minor incidents.
- Provision of roadside emergency call boxes at emergency pullouts.
- Special lighting to enhance visibility of the shoulder.
- Advanced incident detection capabilities.
- Comprehensive incident management program.
- Connection to a traffic management center that serves as the focal point, and
- Dynamic message signs to provide guide sign information and regulatory signs to adapt to the addition of the shoulder as a travel lane.

3.3.3 Queue warning

Queue warning message displays should be implemented at regular intervals to warn of the presence of upstream queues based on dynamic traffic detection. Specific elements of the operational strategy should include the following (Mirshahiet et al., 2007):

- Deployment in conjunction with speed harmonization.
- Sufficient sensor deployment for traffic monitoring to support the strategy.
- Adequate installation of sign gantries to ensure that at least one queue warning sign is in sight at all times.
- An expert system that deploys the strategy based on prevailing roadway conditions without requiring operator intervention. It is critical that this expert system be reliable and accurate to gain the trust and acceptance of the public.
- Uniform signing to indicate congestion ahead, and
- Connection to a traffic management center that serves as the focal point for the system.

3.3.4 Dynamic merge control

At merging points from major interchange ramps, consideration should be given to dynamically metering or closing specific upstream lanes, depending on traffic demand. This could incorporate existing ramp metering systems and offer the potential of delaying the onset of main lane congestion and balancing demands between upstream roadways. Specific elements of the operational strategy should include the following (Mirshahiet et al., 2007):

- An expert system that deploys the strategy based on prevailing roadway conditions without requiring operator intervention. It is critical that this expert system be reliable and accurate to gain the trust and acceptance of the public.
- CCTV cameras to support the monitoring of the system.
- Installation of lane control signals over the main lanes and the ramp lanes with a signal over each travel lane.
- Adequate installation of sign gantries upstream of the deployment to ensure sufficient advance warning is provided to roadway users through the use of dynamic message signs.
- Adequate installation of sign gantries with dynamic message signs upstream of the deployment to provide guide sign information and regulatory signs to adapt to the changes in lane use.
- Uniform signing to indicate merge control is in use.
- Automated enforcement to deter violations.
- A bypass lane for emergency vehicles, transit, or other identified exempt users, and
- Connection to a traffic management center that serves as the system's focal point.

4. Birmingham case study

4.1 Methodology

Drawing from the European experience and with input from the few available U.S. studies, a study procedure was developed and implemented in order to assess the feasibility of temporary shoulder lane use as a strategy to reduce congestion in the Birmingham region. This included:

- Identification of candidate corridors and preliminary assessment of implementation potential.
- Quantitative evaluation of operational impacts from implementation, and
- Estimation of benefits and costs.

Preliminary assessment of temporary shoulder lane implementation along candidate corridors considered traffic demand, level of service (LOS), physical characteristics and geometric restrictions. The assessment led to the selection of a segment of I-65 traversing through the city of Birmingham as a high-priority corridor for further analysis. I-65 is a major North-South interstate freeway that extends from Gary, IN on the north to Mobile, AL on the south.

Simulation modeling was undertaken to analyze the impacts of a temporary shoulder lane use system on a subsection of the I-65 corridor extending from I-459 to University Blvd, which corresponds to the portion of the study corridor near Birmingham that experiences the worst Level-of-Service (LOS). The microscopic simulation model CORSIM was used to perform the analysis.

CORSIM is one of the tools available within TSIS, a suite of simulation models developed by FHWA and used extensively by transportation agencies and practitioners in the U.S. and abroad for over three decades. The CORSIM simulator in TSIS can simulate traffic operations on integrated networks containing freeway and surface streets. The model has the ability to simulate fairly complex geometric conditions and realistic driver behavior after it is appropriately calibrated and validated. Moreover, the model offers the capability to analyze a variety of lane management strategies, a feature of importance for this case study (Sisiopiku & Cavusoglu, 2008).

The Birmingham case study considered the potential use of a shoulder lane in response to both recurring and non-recurring congestion. Key measures of effectiveness (MOEs) and resulting improvements in operational efficiency were obtained for several scenarios and used to assess operational impacts and determine the feasibility of implementation of the proposed strategy.

Quantification of expected benefits and costs from deployment of temporary shoulder lanes along the I-65 corridor in Birmingham was also performed to estimate economic impacts from possible deployment and determine the most economically efficient investment alternative.

The cost-benefit analysis considered life-cycle costs and life-cycle benefits of the project alternatives under study. The life-cycle costs include engineering, construction, and maintenance. Life-cycle benefits include savings in vehicle operation and travel time, safety, and emission reduction. Following the analysis, the costs and benefits were discounted on year-to-year basis and projected for the analysis period 2010 to 2020. A description of the study site characteristics and the scenarios tested follows.

4.2 Study site characteristics

The site chosen to examine the implementation of temporary shoulder lane usage in this research is the northbound I-65 from the junction with I-459 to the University Blvd junction. The geometric characteristics of the selected segment are summarized in Table 2.

According to the Mobility Matters Project in Birmingham, the 2005 average daily traffic volumes along the study segment ranged from 111,000 and 146,000 vehicles per day (vpd). This number is expected to reach 179,000 and 221,000 vpd by 2030 (PBS&J, 2009).

Segment	Number of Lanes	Lane Width (ft)	Shoulder Lane Width (ft)	
			Left	Right
I-459 to US 31	4	12.0	9.5	12.5
US 31 to Alford Ave	3	12.0	12.0	12.5
Alford Ave to Lakeshore Dr	3	12.0	20.0	12.0
Lakeshore Dr to Oxmoor Rd	3	11.5	7.0	12.0
Oxmoor Rd to Green Springs	3	12.0	13.0	8.0
Green Springs to University Ave	3	12.0	10.0	11.5

Table 2. Geometric Characteristics of Study Corridor.

Based on the hourly traffic volumes collected by by the Alabama Department of Transportation (ALDOT) along the I-65, the morning peak hour (which affects primarily the northbound direction) is more critical than the evening peak hour that primarily affects southbound traffic, and thus was selected for further analysis (Sisiopiku et al., 2009).

4.3 Alternatives analysis

Study scenarios were developed and tested with the TSIS simulation model for a period of four hours (5:30 AM to 9:30 AM), including the morning peak hour (6:30 AM to 8:30 AM). The scenarios aimed at examining traffic operations along the study corridor with and without the use of temporary shoulder lanes under normal and incident traffic conditions. More specifically, four scenarios were developed assuming non-incident (normal) traffic conditions to examine the efficiency of using temporary shoulder lanes to ease recurrent traffic congestion on the northbound I-65 corridor. In all four scenarios, the free flow speed was set to 60 mph.

Scenario 1 served as a baseline for comparisons and assumed that the network operates under normal conditions without the use of the shoulder lanes.

Scenario 2 simulated the network with the utilization of the left shoulder as an additional lane from U.S. 31 to the end of the network. The shoulder lane was open during the entire simulation period and represented the case of an added lane.

Scenario 3 was similar to the second study scenario, however, the temporary shoulder lane was available for use only in the morning peak hour (between 6:30 AM and 8:30 AM). This scenario serves as a typical example of an ATM application where the temporary shoulder lane is used in response to congestion.

Scenario 4 simulated the network under normal conditions while opening a small portion (i.e., 600 feet) of the right shoulder upstream of three study exits, namely Alford Ave (Exit 254), Lakeshore Pkwy (Exit 255), and Oxmoor Rd (Exit 256) for the total simulation time. This scenario tested the possibility of using the right shoulder as an additional exit lane in order to minimize the potential impact that long queues of exiting vehicles may have on traffic operations along the mainline.

Moreover, three scenarios were developed to examine the efficiency of using temporary shoulder lanes under incident conditions. For practical purposes these were numbered consequently as Scenarios 5 through 7. All incident scenarios assumed that an incident occurred blocking the right lane of link (564,565) for 1 hr (i.e., from 6:30 AM to 7:30 AM). The incident site was located roughly in the middle section of the study network.

Scenario 5 considered the presence of the incident and assumed that no actions were taken. This scenario serves as the incident case base line for comparison purposes (i.e., do-nothing approach).

Scenario 6 simulated the network with the incident presence but assumed that vehicles were allowed to ride on the left shoulder lane downstream of the incident for one hour (6:30 AM to 7:30 AM) to minimize the impact of the incident on traffic operations.

The last scenario considered (**Scenario 7**) is similar to Scenario 6, except for the fact that the left shoulder lane downstream of the incident is open for use for 2 hours i.e., from 6:30 AM to 8:30 AM, to further expedite the clearance of the incident and return to normal operations. Table 3 provides a summary of scenarios considered in the Birmingham case study.

4.4 Results

4.4.1 Operational analysis results

The results presented in Table 4 are for the entire study corridor (i.e., network wide statistics) for the seven scenarios considered in the study. MOEs considered include the total travel time (hours), the total delay time (hours), the average travel speed (mile/hour), the delay time (hours), and the total time (hours). Analysis and interpretation of the results under normal - and incident conditions follows.

Scenarios	Description	Available lanes	Left shoulder lane in use (Duration)	Right shoulder lane in use (Duration)
Normal Traffic Conditions				
1	No shoulder use	3	0 (5:30 AM to 9:30 AM)	0 (5:30 AM to 9:30 AM)
2	Continuous left shoulder lane use	3	1 (5:30 AM to 9:30 AM)	0 (5:30 AM to 9:30 AM)
3	Continuous left temporary shoulder lane use	3	1 (6:30 AM to 8:30 AM)	0 (5:30 AM to 9:30 AM)
4	Right shoulder lane use at exits	3	0 (5:30 AM to 9:30 AM)	1 (5:30 AM to 9:30 AM)
Incident Conditions				
5	No shoulder use	3 or 2	0 (5:30 AM to 9:30 AM)	
6	Continuous left temporary shoulder lane use down-stream of the incident	3 or 2	1 (6:30 AM to 7:30 AM)	
7	Continuous left temporary shoulder lane use downstream the incident	3 or 2	1 (6:30 AM to 8:30 AM)	

Table 3. Summary of Case Study Scenarios.

Scenario	1	2	3	4	5	6	7
Shoulder Lane Use	No	Left	Left; Peak only	Right; 3 exits	No	Left; 1hr	Left; 2 hrs
Total Travel Time (hrs)	6,790	3,963	4,478	6,523	7,872	6,758	5,498
Total Delay Time (hrs)	3,394	446	991	3,127	4,598	3,337	1,963
Avg. Travel Speed (mph)	29.90	53.10	46.60	31.30	25.10	30.30	38.70
Delay Time (min/mile)	1.01	0.13	0.28	0.92	1.42	0.98	0.57
Total Time (min/mile)	2.01	1.13	1.29	1.93	2.43	1.99	1.57

Table 4. Network-Wide Results for All Scenarios; Birmingham, AL Case Study.

Scenarios with normal traffic conditions

According to the findings in Table 4, the use of the left shoulder lane in Scenarios 2 and 3 results in considerable savings in travel time and delays, as compared to the baseline (Scenario 1). As expected, the continuous availability of an extra lane (Scenario 2) results in the largest improvements, slashing total time by 42% (from 6,790 hrs in Scenario 1 to 3,963 hrs in Scenario 2).

The temporary use of the NB shoulder lane for 2 hrs during the morning peak (Scenario 3) still shows a significant improvement over current conditions resulting in a reduction in the total network travel time by 34% and delay by 71% when compared to the baseline (Scenario 1). Similar gains are observed in average speed where the 29.9 mph average network speed observed under regular conditions (Scenario 1) increases by 56% (to 46.6 mph) under the ATM operations, i.e., when the left shoulder lane is open during the peak hour from 6:30 AM to 8:30 AM.

The results clearly indicate the excellent potential of temporary shoulder lane use as an ATM tool for addressing recurrent congestion along I-65. On the other hand, the use of the right northbound shoulder lane upstream of three exit locations (Scenario 4) shows only a small positive impact and results in a small reduction in the total network travel (4%) time and delay (9%) over Scenario 1. A slight increase (5%) in average speed was also noticed (from 29.9 mph to 31.3 mph) in this case.

When comparing the two temporary shoulder lane options, i.e., continuous left lane shoulder versus right shoulder near exits, the former is clearly a winner, as the anticipated benefits clearly overshadow those expected from its short length, temporary, right shoulder use counterpart.

Scenarios with incident conditions

As anticipated, an incident blocking one general purpose traffic lane for 1 hr (Scenario 5) further degraded the overall performance of the study network. Compared to non-incident conditions (Scenario 1), the do-nothing approach under incident conditions resulted in a decrease in average speed of 16% (from 29.9 mph to 25.1 mph) and average delay time increase of 41% (from 1.01 min/mile to 1.42 min/mile).

In Scenario 6 the left shoulder lane downstream of the incident site was opened just for one hour following the onset of the incident. For this scenario, the simulation results show that the average network speed increased by 21% (from 25.1 mph to 30.3 mph), and the average delay time decreased by 31%, as compared to Scenario 5. These gains resulted in network performance comparable to non-incident conditions. As expected, the savings in travel time and delay further increased when the temporary shoulder lane remained in operation for an extra hour following the incident removal (Scenario 7).

Overall, the network-wide results from the incident case study demonstrate the great potential operational benefits of the temporary shoulder lane use as a traffic management strategy in case of an incident.

4.4.2 Cost-benefit analysis results

A detailed cost-benefit analysis was performed to estimate economic impacts from possible deployment of temporary shoulder lane use strategies along the study segment of the I-65 freeway in Birmingham. The cost-benefit analysis compared anticipated costs and benefits from each of the study scenarios to the base case scenario (Scenario 1) in order to find the most cost-effective method.

From the CORSIM outputs for the different scenarios presented earlier, the fuel consumption, vehicle miles of travel, travel time, and emissions outputs were obtained and used to calculate Vehicle Operating Costs (VOC), value of time (VOT), accident costs, and emission costs. These costs were utilized to calculate and compare the benefit-cost ratios for different scenarios on an annual basis. Details are available in Sisiopiku et al, 2009b.

The benefit components for different scenarios are summarized in Table 5 and the resulting benefit-cost ratios for different scenarios are calculated and presented in Table 6. Table 5 shows clearly that VOC, VOT, accident costs and emission costs are lower when shoulder lane use is permitted. Furthermore, Table 6 indicates that the temporary use of the left shoulder lane along the I-65 study section during the 2 hour morning peak (Scenario 3) is expected to result in $12.6M in savings. Further savings can be realized by utilizing the shoulder lane for more extended periods of time (such as in Scenario 2). For alleviation of traffic congestion due to incidents, the most cost-effective option studied is provided by Scenario 7. The findings of benefit-cost analysis are in close agreement with those of the traffic impact analysis presented earlier.

Shoulder Lane Operation- Normal Conditions				
Scenario	VOC [Mil $/year]	VOT [Mil $/year]	Accident Cost [Mil $/year]	Emission Cost [Mil $/year]
1	7.607	34.213	11.226	0.0043
2	5.881	19.964	11.645	0.0041
3	6.304	22.563	11.538	0.0042
4	7.519	32.867	11.226	0.0043
Shoulder Lane Operation- Incident Conditions				
5	7.975	39.662	10.826	0.0044
6	7.632	34.719	11.173	0.0043
7	6.880	27.498	11.557	0.0042

Table 5. Benefit Components for Different Scenarios.

Shoulder Lane Operation- Normal Condition				
Scenario	Total Costs [M $/year]	Total Benefit Components [M $/year]	Total Benefits with respect to Base Case [M$/year]	B/C Ratios
1 (Base)	0.043	53.050	-	
2	1.122	37.494	15.556	13.87
3	1.122	40.409	12.641	11.27
4	1.434	51.613	1.434	3.75
Shoulder Lane Operation- Incident Conditions				
5 (Base)	0.043	58.467	-	-
6	1.108	53.528	4.939	4.46
7	1.108	45.939	12.528	11.31

Table 6. Benefit/Cost Ratios for Different Scenarios.

4.5 Case study summary findings

Simulation analysis was performed to quantify the potential benefits of a temporary left shoulder lane use system on a segment of I-65 in Birmingham in response to recurrent and non-recurrent congestion. The results from the simulation analysis, coupled with findings from a cost-benefit analysis, were used to demonstrate the potential of the strategy to improve traffic operations and justify the need for deployment of the proposed strategy at the study location.

It was found that the use of temporary shoulder lanes can have a very positive impact on traffic operations along I-65 when implemented in response to both recurrent- and/or non-recurrent congestion. In this study, the temporary use of the left northbound shoulder lane for 2 hours during the morning peak (Scenario 3) resulted in a reduction in the total network travel time by 34% and delay by 71% compared to current operations (Scenario 1).

The use of right shoulder lanes upstream of exit ramps tested in this study provide some relief but had far less impact on network performance, compared to the continuous left shoulder lane usage. These results clearly indicate the tremendous potential of temporary shoulder lane use as an active traffic management tool for addressing recurrent congestion along I-65.

It was also found that considerable improvements in traffic operations can be achieved by utilizing the temporary shoulder lane downstream of an incident as an ATM measure. In the Birmingham study and under incident conditions, the utilization of the temporary shoulder lane resulted in an increase in average network speed by 21% and a decrease in average delay time by 31%, as compared to the do-nothing approach. These gains are significant and provide further proof of the potential of temporary shoulder lane use within the ATM context as a tool for incident management.

The results from the benefit-cost analysis provide further justification for the use on temporary shoulder lanes. It can be seen that the total benefits from implementation of this strategy outweigh the total costs, which further confirms that the temporary shoulder lane use treatment is an economically viable solution both in the short and long terms.

5. Conclusions and recommendations

Overall, ATM seeks to introduce new congestion management strategies to the U.S. while enhancing the effectiveness of existing strategies. It should be viewed as the next logical step in the evolution of congestion management in this country rather than a radical change from previous practice. The European experience with ATM clearly demonstrates its positive impacts on traffic operations and safety and thereby, its tremendous potential for alleviating traffic congestion in the US.

However, the implementation of ATM is a significant investment so the potential benefits would have to be clearly defined and sufficient to justify the costs. To better assess potential costs and gains, careful screening of candidate test sites should take place first, followed by detailed assessment of operational and cost impacts from implementation. This practice will help identify opportunities and impediments from implementation, and document technology, policy, and other needs.

The success of implementation of ATM greatly depends on public support for the project and positive public perception. Thus, the role of public education in the early planning stage is critical and should not be overlooked. Focus groups, open public discussion forums, public information sessions, and media coverage are useful tools that can assist local agencies to obtain input from the public and other local stakeholders and educate the road users about their rights and responsibilities as they use the new ATM systems.

6. Acknowledgment

The authors gratefully acknowledge the Urban Transportation Center for Alabama for the financial support for this research. Moreover, the contributions of Mrs. Germin Fadel, Ms. Ozge Cavusoglu, Mr. Andy Sullivan and Dr. Saiyid Sikder to the Birmingham case study are greatly appreciated.

7. References

Berman, W., Differt, D. H., Aufschneider, K., DeCorla-Souza, P., Flemer, A., Hoang, L. T., Hull R., Schreffler E., and Zammit, G. (2006). *Managing Travel Demand: Applying European Perspectives to U.S. Practice.* Washington, DC: Federal Highway Administration.

FHWA (2003). *Freeway Management and Operations Handbook -Roadway Improvements.* Office of Transportation Management. Washington, D.C.

FHWA (1994). *IVHS Institutional Issues and Case Studies.* U.S. Department of Transportation. Cambridge, MA: Federal Highway Administration.

MnDOT (2010). "Minnesota's Smart Lanes Go Live July 29", Retrieved May, 2011, from Minnesota Department of Transportation: http://www.dot.state.mn.us/metro/news/10/07/28smartlanes.html

Mirshahiet, M., Obenberge, J., Fuhs, C.A., Howard, C.E., Krammes, R.A., & Kuhn, B. T. (2007). *Active Traffic Management: The Next Step in Congestion Management.* Washington, DC: Federal Highway Administration, FHWA-PL-07-012.

NCDOT (n.d.). *High Occupancy Vehicle (HOV).* Retrieved December 28, 2008, from North Carolina Department of Transportation: http://www.ncdot.org/projects/HOV/

PBS&J. (2009). *Tier 1 Alternatives Evaluation:I-65 / US 31 Mobility Matters Project.* PBS&J Team Project.

Sisiopiku, V.P., Cavusoglu, O., and Fadel, G. (2009). Active Traffic Management Opportunites and Challenges for Implementaion: ITE Technical Confrerence and Exhibit Compendium of Technical Papers, Phoenix, AZ.

Sisiopiku, V.P., Sullivan, A., Fadel, G., and Sikder, S. (2009b). *Implementing Active Traffic Management Strategies in the U.S.* Final Report to the University Transportation Center for Alabama.

Sisiopiku, V.P., and Cavusoglu, O. (2008). Operational Impacts from Managed Lanes Implementation in Birmingham: ITE Annual Meeting and Exhibit Compendium of Technical Papers, Anaheim, CA.

Stone, C., Hammond, P., and Lenzi, J. (2007). *Active Traffic Management: The Next Step in Congestion Management.* Washington: Washington State Department of Transportation.

Tignor, S. C., Brown, L. L., Butner, J. L., Cunard, R., Davis, S. C., Hawkins, H. G., et al. (1999). *Innovative Traffic Control Technology and Practice in Europe.* Washington, D.C.: Federal Highway Administration.

8

How to Provide Accurate and Robust Traffic Forecasts Practically?

Yang Zhang
Shanghai Municipal Transportation Information Center
China

1. Introduction

With the development of our modern cities, growing traffic problems adversely affect people's traveling convenience more and more, which has become one of the most crucial factors considered in urban planning and design in recent years. Urban traffic congestion is a severe problem that significantly reduces the quality of life in particularly metropolitan areas. However, frequently constructing new roads is not realistic and untenable in social and economic aspects. In the effort to deal with this intractable problem, so-called intelligent transportation systems (ITS) technologies are successfully implemented widely throughout the world nowadays. ITS with two important components advanced traffic management systems (ATMS) and advanced traveler information systems (ATIS) aim to relieve the increasing congestion and decrease travel time through providing information to the drivers by means of radio broadcasts or dynamic route guidance systems.

The provision of accurate real-time information and predictions of traffic states such as traffic flow, travel time, occupancies, etc., is much fundamental and contributive to the great success of ITS (Chen et al., 2010; Dong et al., 2010; Vlahogianni et al., 2004; Lam et al., 2006; Tan et al., 2009; Tang et al., 2003; Thomas et al., 2010; Zhang & Liu, 2008, 2009c). As an important part of ITS, traffic states analysis and traffic forecasting are important in directing commuters to pick optimal routes, which have attracted many researchers to focus on this subject in recent decades. In general, as illustrated in the statement, the traffic forecasting "can be separated into two paradigms: the empirical based, incorporating fairly standard statistical methodology on the one hand, and that based on traffic process theory, either of demand or of supply, on the other" (Van Arem et al., 1997).

Because of the feasibility of data collection from numerous kinds of equipments and the requirements of dynamic management, the empirical approaches for traffic forecasting correspond with the development trends of ITS. It aims to find out the hidden regularity of traffic states through the random and uncertain traffic data by systematic analysis and a variety of mathematics/physics methods.

The empirical approaches can be approximately divided into two types: basic forecasts approaches and combined forecasts approaches. The basic forecasts approach means to predict the traffic state using a certain particular prediction model. The robustness and accuracy of these approaches lie on the prediction models themselves. Furthermore, basic forecasts approaches can be roughly classified into two types: parametric and nonparametric techniques. Both techniques have shown their own advantages on different occasions in recent years (Tsekeris & Stathopoulos, 2010; Zhang & Liu, 2009f, Zhang & Liu, 2010; Zhang et al., 2010). On the basis of the classification, the chapter provides a systematic review of these models such as historical-mean (HM), filtering algorithm, linear and nonlinear regression, autoregressive process, neural network (NN), fuzzy systems, support vector regression (SVR), and Bayesian networks, etc.

The combined forecasts approach means to combine different forecasts into a single one that is assumed to produce a more accurate forecast. The robustness and accuracy of combined forecast approaches lie not only on the prediction effect of individual prediction model, but also on the efficiency of combination. Because the combined method is to apply each predictor's unique feature to capture different patterns in the data, it would give a smaller error variance than any of the individual methods (Bates & Granger, 1969). This advantage may make the approach fully scalable to the very large amounts of traffic data practically. Due to its simplicity and practicability, the combined forecasts approach becomes very important to traffic forecasting, and researchers have focused on it, both theoretical and applied.

Though the data-driven traffic forecasting gains many achievements, there still exist some unsolved problems. From the practical point of view, data gained from some detectors are incomplete, i.e., partially or completely missing or substantially contaminated by noises. The missing data sometimes render an entire dataset useless, which is a major hurdle in analyzing traffic information. As missing data treatment is an important preparation step for effective management of ITS, some proper solutions to solve missing data problems are provided in the chapter. And the chapter ends with a brief introdcution of Shanghai Integrated Transportation Information Platform (SITIC), which represents the level of informatization development in transportation.

2. A brief review of data-driven traffic forecasting

The data-driven traffic forecasting refers to predicting the future state of a certain transportation system based on the historical data, existing traffic data and the related statistics data (Brockwell & Davis, 2002; Chrobok, 2004). Traffic forecasting is a branch of forecasting, and it is an important part of modern transportation planning and intelligent transportation system. Usually, traffic flow, average speed and travel time etc., are defined as the basic parameters of traffic state. Specifically, traffic forecasting is essentially the prediction of these basic parameters based on dynamic road traffic time series data. For instance, most of literature foucs on traffc flow forecasting (Jiang & Adeli, 2004; Qiao et al., 2001; Abdulhai et al., 1999; Castillo et al., 2008; Chen & Chen, 2007; Dimitriou et al., 2008; Ding et al., 2002; Huang & Sadek, 2009; Ghosh et al., 2005, 2007; Smith et al., 2002), travel time forecasting, and related analysis such as validation, optimization, etc. (Chan et al., 2003; Chan & Lam, 2005; Chang et al., 2010; Kwon, 2000; Kwon & Petty, 2005; Lam, 2008; Lam et al., 2002, 2005, 2008; Lam & Chan, 2004; Lee et al. 2009; Nath et al., 2010; Schadschneider et al., 2005; Tam & Lam, 2009; Tang & Lam, 2001; Yang et al., 2010).

Overall, the process of traffic state variation is a real-time, nonlinear, high dimensional and non-stationary stochastic process. With the shortening of statistical time range, the stochastic and uncertainty of traffic state are more and more strong. Short-term traffic state variation is not only related to the state of the local road section over the past few hours, but also influenced by the traffic states of upstream and downstream road sections, weather situation and unexpected events, etc.

From the spatial and temporal point of view, the traffic state can reflect regular variation. For example, the traffic states of various road sections of urban road network during peak and non-peak period show periodic variation respectively; and the traffic states in urban highway traffic on weekdays and weekends also show different periodic variation, which reflects the temporal regularity of road network traffic. Meanwhile, the urban road network topology, the length of each road link, lane width and traffic direction, etc. can determine the variation of traffic state on a particular road link, which reflects the spatial regularity of road network traffic. Therefore, in the research of transportation prediction, it is essential to fully consider real-time traffic state variation with the randomness and regularity temporally and spatially. Namely, real-time traffic forecasting should predict the future traffic state on the basis of studying the specific sections of the historical traffic data, the whole spatial-temporal road network traffic condition variation, weather situation, and other influence factors. Fig.1 describes the framework of data-driven traffic forecasting models.

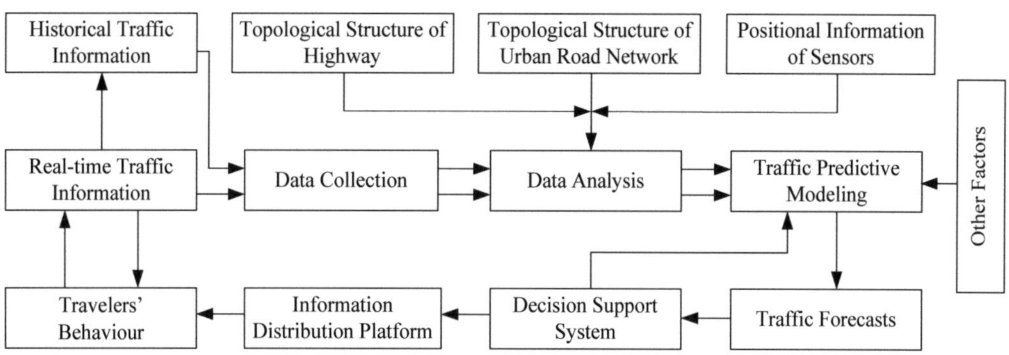

Fig. 1. The framework of traffic forecasting models.

3. Traffic forecasting approaches

The following factors are usually used for the classification of traffic forecasting approaches: single road link or transportation network, freeways or urban streets, physical models or mathematical methodologies, univariate or multivariate method, etc. From the methodology point of view, the traffic forecasting approaches can be divided into two types: the empirical based approaches and traffic process theory based approaches. For the convenience of data collection from numerous kinds of equipments, a large amount of the historical traffic information and real-time traffic information can be obtained. And the empirical approaches become the new trend of ITS. In this part, we focus on the achievements concerned with empirical approach according to its classification.

3.1 Basic forecasts approaches

A large amount of scientific literature has been concerned with basic forecasts approaches. On the basis of the classification, the chapter provides a systematic review of parametric and nonparametric traffic forecasting techniques briefly.

3.1.1 Parametric traffic forecasting approaches

Since the early 1980s, extensive variety of parametric approaches has been employed ranging from historical average algorithms (Smith & Demetsky, 1997; Wu et al., 2004), smoothing techniques (Smith & Demetsky, 1997; Williams et al., 1998), linear and nonlinear regression (Deng et al., 2009; Lu et al., 2009; Zhang & Rice 2003; Sun et al., 2003), filtering techniques (Ross, 1982; Okutani & Stephanedes, 1984; Whittaker et al., 1997; Chien & Kuchipudi, 2003; Stathopoulos & Karlaftis, 2003), to autoregressive linear processes (Min et al., 2010; Min & Wynter, 2011). Thereinto, the autoregressive integrated moving average (ARIMA) (Ahmed & Cook, 1979) family of models such as simple ARIMA (Levin & Tsao, 1980; Nihan & Holmesland, 1980; Hamed et al., 1995; Smith, 1995; Williams, 1999), ATHENA (Kirby et al., 1997), subset ARIMA (Lee & Fambro, 1999), SARIMA family (Smith et al., 2002; Williams et al., 1998, 2003; Ghosh et al., 2005), are classical milestones in forecasting area. Such time series methods belong to time domain approaches, and frequency domain approaches like spectral analysis, "which are regressions on periodic sines and cosines, show their important insights into traffic data which may not apparent in an analysis in the time domain only" (Stathopoulos & Karlaftis, 2001a, b). The parametric traffic forecasting approach is the milestone of the traditional time series forecasting. And it brings significant developments for traffic forecasting.

3.1.2 Nonparametric traffic forecasting approaches

Lately extraordinary development of distinct nonparametric techniques, including nonparametric regression, neural networks, etc., has shown that they may be able to become a high potential alternative to their parametric counterparts (Huisken, 2003; Lam et al., 2006). In essence, nonparametric statistical regression can be regarded as a dynamic clustering model that relies on the relationship between dependent and independent traffic variables. (Davis & Nihan, 1991; Smith & Demetsky, 1997; You & Kim, 2000; Smith et al., 2000, 2002; Clark, 2003; Turochy, 2006). In other words, it attempts to identify past information that are similar to the state at prediction time, which leads to easily implemented nature. Over the past decade, another nonparametric technique, artificial neural networks (ANNs) have been applied in traffic forecasting because of their strong ability to capture the indeterministic and complex nonlinearity of time series (Smith & Demetsky, 1994, 1997; Chang & Su, 1995; Dougherty & Cobbet, 1997; Lam & Xu, 2000; Park et al., 1999; Dharia & Adeli, 2003; Wei et al., 2009; Wei & Lee 2007; Lee, 2009). Motivated by the universal approximation property, neural network models ranging from purely static to highly dynamic structures include the multilayer perceptrons (MLPs) (Clark et al., 1993; Vythoulkas, 1993; Lee & Fambro, 1999; Gilmore & Abe, 1995; Ledoux, 1997; Innamaa, 2000; Florio & Mussone, 1996; Yun et al., 1998; Zhang, 2000; Chen et al., 2001), the radial basis function (RBF) ANNs (Lyons et al., 1996; Park et al., 1998; Park & Rilett, 1998; Chen et al., 2001), the time-delayed ANNs (Lingras et al., 2000; Lingras & Mountford, 2001; Yun et al., 1998; Yasdi 1999; Abdulhai et al., 1999; Dia, 2001; Ishak & Alecsandru, 2003), the recurrent

ANNs (Dia, 2001; Van Lint et al., 2002, 2005), and the hybrid ANNs (Abdulhai et al., 1999; Chen et al., 2001; Lingras & Mountford, 2001; Park, 2002; Yin et al., 2002; Vlahogianni et al., 2005; Jiang & Adeli, 2005; Quek et al., 2006), etc. Besides the above neural networks models, computational intelligence (CI) techniques that encompass fuzzy systems, machine learning and evolutionary computation have been successfully developed in the field of traffic forecasting. For instance, some literature applies Bayesian networks (Zhang et al. , 2004; Castillo et al., 2008) and Bayesian inference based regression techniques (Khan, 2011; Tebaldi et al., 2002; Sun et al., 2005, 2006; Zheng et al., 2006; Ghosh et al., 2007), some literature uses fuzzy systems or fuzzy NNs to predict the traffic states (Dimitriou et al., 2008; Quek et al., 2009). While others start to explore support vector regression (SVR) to model traffic characteristics and produce prediction of traffic states (Castro-Neto, 2009; Ding et al., 2002; Hong, 2011; Hong et al., 2011; Wu et al., 2004; Vanajakshi & Rilett, 2004). The recent application of different CI techniques and hybrid intelligent systems has shown that the rapidly expanding research field is promising.

3.2 Combined forecasts approaches

The basic idea of the combined forecasts approach is to apply each predictor's unique feature to capture different patterns in the data (Zhang & Liu, 2009d, 2009e). The complement in capturing patterns of data sets is theoretically essential for more accurate prediction (Timmermann, 2005; Huang, 2007). "Both theoretical and empirical findings suggest that combining different methods can be an effective way to improve forecast performances." (Yu et al., 2005a). The linear combining forecasts methodology has a long historical background. Compared to computational intelligence based nonlinear ensemble forecasting models (Chen & Zhang, 2005; Chen & Chen, 2007), the linear combination retains the conceptual and computational simplicity. In the part, we focus on the application of linear combination method. Researchers have proposed various combined methods since the pioneering work of Bates and Granger. Clemen provided a review and annotated bibliography of the literature for reference (Clemen, 1989). "Research in various fields has strongly suggested that the performance of prediction can be enhanced when (sometimes even in simple fashion) forecasts are combined." (Yang, 2004).

Basically, we can describe the main problem of combined forecasts as follows. Suppose there are N forecasts such as $\hat{V}_{P1}(t)$, $\hat{V}_{P2}(t)$, …, $\hat{V}_{PN}(t)$ (including correlated or uncorrelated forecasts), where $\hat{V}_{Pi}(t)$ represents the forecasting result obtained from the ith model during the time interval t. The combination of the different forecasts into a single forecast $\hat{V}_P(t)$ is assumed to produce a more accurate forecast. The general form of such a combined forecast can be described with formula

$$\hat{V}_P(t) = \sum_{i=1}^{N} w_i \hat{V}_{Pi}(t) \tag{1}$$

where w_i denotes the assigned weight of $\hat{V}_{Pi}(t)$, and commonly the sum of the weights is equal to one, i.e., $\sum_i w_i=1$. Our studies mainly investigate the combined models with the additional restriction that no individual weight can be outside the interval [0, 1]. Various methods can be applied to determine the weights used in the combined forecasts. Four common methods are presented in the following.

3.2.1 Equal Weights (EW) methods

The EW method, applying a simple arithmetic average of the individual forecasts, is a relatively robust method with low computational efforts. Namely, each w_i is equal to 1/N (i=1, 2, …, N), where N is the number of forecasts. The beauty of using the simple average is that it is easy to understand and implement, not requiring any estimation of weights or other parameters (Jose & Winkler, 2008). This makes it robust because they are not sensitive to estimation errors, which can sometimes be substantial. It often provides better results than more complicated and sophisticated combining models (Clemen, 1989). Although the approach has non-optimal weights, it may give rise to better results than time-varying weights that are sometimes adversely affected by some unsystematic changes over time. Under the circumstances, the method has the virtues of impartiality, robustness and a good "track-record" in time series forecasting. It has been consistently the choice of many researchers in the combination of forecasts.

3.2.2 Optimal Weights (OW) methods

Bates & Granger proposed that using a MV criterion can determine the weights to adequately apply the additional information hidden in the discarded forecast(s) (Bates & Granger, 1969), and Dickinson extended the method to the combinations of N forecasts (Dickinson, 1973). Assuming that the individual forecast errors are unbiased, we can calculate the vector of weights to minimize the error variance of the combination according to the formula

$$\boldsymbol{w} = \boldsymbol{M}_V^{-1}\mathbf{I}_n(\mathbf{I}_n'\boldsymbol{M}_V^{-1}\mathbf{I}_n)^{-1} \tag{2}$$

where $\mathbf{I}_n$ is the n×1 matrix with all elements unity (i.e. n×1 unit vector) and $\boldsymbol{M}_V$ is the covariance matrix of forecast errors (e.g. M_{Vij} is the covariance between the error of forecast i and forecast j at a particular point in time). Granger & Ramanathan pointed that the method is equivalent to a least squares regression in which the constant is suppressed and the weights are constrained to sum to one (Granger & Ramanathan, 1984). In the case of a combination of two forecasts, we suppose there is no correlation between forecast errors.

3.2.3 Minimum Error (ME) methods

The ME method minimizes the forecasting errors when combining individual forecasts into a single one. A solution for this method applies linear programming (LP) whose principle and computational process are described as follows (Yu et al., 2005a). Set the sum of absolute forecasting error (i.e., $\sum_i E_i(t)$ during the time interval t) as

$$F_{LP} = \sum_{i=1}^{N} |E_i(t)| = \sum_{i=1}^{N} \left| w_i(t)\left(\hat{V}_{Pi}(t) - V_O(t)\right)\right| \quad t = 1, 2, \cdots, \mathrm{T} \tag{3}$$

where F_{LP} is the objective function of LP; $V_O(t)$ denotes the observed value during the time interval t and T the number of forecasting periods. To eliminate the absolute sign of the objective function, assume that

$$u_i(t) = \frac{|E_i(t)| + E_i(t)}{2} = \begin{cases} E_i(t), & E_i(t) \geq 0 \\ 0, & E_i(t) < 0 \end{cases} ,$$

$$v_i(t) = \frac{|E_i(t)| - E_i(t)}{2} = \begin{cases} 0, & E_i(t) \ge 0 \\ -E_i(t) & E_i(t) < 0 \end{cases} \tag{4}$$

The introduction of $u_i(t)$ and $v_i(t)$ aims to transform the absolute sign of the objective function so as to be consistent with the standard form of LP. Obviously, $|e_i(t)|=u_i(t)+v_i(t)$, $e_i(t)=u_i(t)-v_i(t)$. On the basis of the above specification, the LP model can be constructed as follows:

$$\begin{cases} Min\ O = \sum_{i=1}^{N} (u_i(t) + v_i(t)), \\ \sum_{i=1}^{N} w_i(t)\left(\hat{V}_{Pi}(t) - V_O(t)\right) - u_i(t) + v_i(t) = 0, \\ \sum_{i=1}^{N} w_i(t) = 1, \\ w_i(t) \ge 0,\ u_i(t) \ge 0,\ v_i(t) \ge 0,\ i = 1, 2, \cdots, N,\ t = 1, 2, \cdots, T \end{cases} \tag{5}$$

where i denotes the number of individual forecasts, and t represents the forecasting periods. In the equation group, assuming $w_i \ge 0$ aims to make every forecast method contribute to the combined forecasting results. The ME method is equivalent to a simple dynamic linear programming problem; thus, the optimal solutions to the LP can be obtained by the simplex algorithm. The method is an effective combination methodology with time-variant weights.

3.2.4 Minimum Variance (MV) methods

The linear combining forecasts methodology has a long historical background. Researchers

Since the negative value of the weight has no factual meaning, researchers usually add the restriction that no individual weight can be outside the interval [0, 1] practically. The main ideas are described as

$$\begin{cases} Min\ (w_i \boldsymbol{M}_V w_i^{\mathrm{T}}), \\ \sum_{i=1}^{N} w_i = 1, \quad i = 1, 2, \cdots, N \\ w_i \ge 0, \end{cases} \tag{6}$$

where $\boldsymbol{M}_V$ is the matrix of error variance. By solving the quadratic programming (QP) problems, an optimal weight set can be obtained for the combining forecasts (Yu et al., 2005b). The problem with this optimizing approach is that it still requires $\boldsymbol{M}_V$ to be properly estimated. Practically, $\boldsymbol{M}_V$ is often not stationary, in which case it is estimated on the basis of a short history of forecasts and thus the method becomes an adaptive approach to combining forecasts (De Menezes et al., 2000).

4. Data imputation

Various imputation techniques have been developed in the past decade. Techniques including naïve imputation, expectation maximization (EM) algorithm (Schafer, 1997; Dempster et al., 1977), data augmentation (DA) algorithm (Tanner & Wong, 1987), and regression imputation, etc. lead logically to modern approaches. Regression imputation and MI have been proved to be more effective than the others, especially the latter one (Ni et al., 2005; Zhong et al., 2004). State space methodology is found to be extremely significant to ensure more accurate results in nearest nonparametric regression (Kamarianakis & Prastakos, 2005). The amelioration including the historical information in the state space may further improve imputation accuracy. Zhang & Liu proposed LS-SVMs method incorporating with the multivariate state space approach to recover missing traffic flow data in arterial streets of Xuhui district, Shanghai (Zhang & Liu, 2009a, 2009b). The state space not only incorporates lagged values but also is supplemented with aggregate measures such as historical information, spatial information, etc. Fully applying spatial and temporal information, the state space based approaches can model the traffic flow successfully. In this part, we focus on the imputation techniques based on state space method (Zhang & Liu, 2009a, 2009b).

In time series, state is defined as a series of system values measured during the past k intervals ($k \in \mathrm{N}$). Measurements at time t, t-1,..., t-k compose a state vector and k is an appropriate number of lags. A state vector of traffic flow measured by loop detector(s) l every F minute(s) can be described by:

$$\boldsymbol{X}_l(t,k_l)=\left[V_l(t),V_l(t-1),\cdots,V_l(t-k_l)\right] \tag{7}$$

where $V_l(t)$ denotes the traffic flow from the detector(s) l during the time interval t; $V_l(t\text{-}1)$ represents the traffic flow from the detector(s) l during the previous F-minute interval, etc. If L loop detectors are considered around the object detector(s) in the traffic network, the values of l range from 1 to L ($L \in \mathrm{N}$). When $t \le k_l$, $\boldsymbol{X}_l(t, k_l)$ contains the last k_l-t+1 parameters measured in theday before the chosen particular day. Object detector(s) can be defined as the detector(s) with missing data.

Considering the historical information in the past week(s), the state space $\boldsymbol{X}(t, L)$ can be defined as:

$$\boldsymbol{X}(t,L)=\left[\boldsymbol{X}_1(t,k_1),\boldsymbol{X}_2(t,k_2),\cdots,\boldsymbol{X}_{\mathrm{L}}(t,k_L),V_{Ghist,w}(t)\right] \tag{8}$$

where $V_{Ghist,w}(t)$ is the historical traffic flow from the object detector(s) at the time-of-day and day-of-the-week associated with time interval t at week w that is usually selected as the previous week. The selection of appropriate L and k_l for each detector l is based on the spatio-temporal analysis of traffic flows collected from loop detectors at different intersections. Input-output pairs for the training process can apply the vectors

$$\{\boldsymbol{X}(t,L),V_{\mathrm{G}}(t)\},\ t\in\left[k_{max},\mathrm{T}\right],\ k_{max}=\max(k_1,k_2,\cdots,k_L) \tag{9}$$

where $V_G(t)$ is the traffic flow obtained from the object detector(s) G during the time interval t; T is the number of time intervals in a day; k_{max} denotes the maximum value among the lags k_l, l=1, 2,..., L. This training process must suppose the good condition of detector(s) G and close relation between $V_G(t)$ and $V_l(t)$. The total number of training samples is (T-k_{max}+1). When detector(s) G cannot supply complete data $V_G(\mathrm{T}+h)$ at time T+h, $h \in \mathrm{N}$, due to some malfunctions, vectors $\boldsymbol{X}(\mathrm{T}+h, L)$ are used as input variables to obtain the predicted results $\widehat{V}_G$ (T+h) that can replace the missing values. Comparison between the imputed values and the actual ones $V_G(\mathrm{T}+h)$ can be utilized to verify the efficiency of different imputation methods.

5. A brief review of Shanghai integrated transportation information platform

In recent years, we have been exploring traffic informatization and building Shanghai Integrated Transportation Information Platform (SITIC) that provides a mechanism to connect isolated islands of information. After three periods of construction, the system software/hardware, backbone networks, information distribution channels have been completed successfully. The guiding thought for the development of SITIC is "Investigating the present state, revealing the objective laws, and guiding the urban transport more scientifically and efficiently".

Classifying the transportation into Road Traffic, Public Traffic, Inter-city Traffic and District/Transport Hub, sorts of information of vehicles and people were collected from kinds of sources, which is the basis of the normal running of SITIC and further data mining. Different real-time information on the transportation of Shanghai can be clearly shown in SITIC. Researching the transportation problems in metropolis, especially the traffic prediction, we found that mastering the situation of transportation is important to traffic management, which leads to the essentiality of level division of the road network into macro (network), meso (district), and micro (link) levels. Meanwhile, data gained from some detectors are incomplete, i.e., partially or completely missing or substantially contaminated by noises. This may be caused by malfunctions in data collection and recording systems that often occur in practice. The missing data sometimes render an entire dataset useless, which is a major hurdle in analyzing traffic information. Missing data treatment is another important preparation step for effective management of intelligent transportation systems (ITS). The following figure describes the contents and function of the platform briefly.

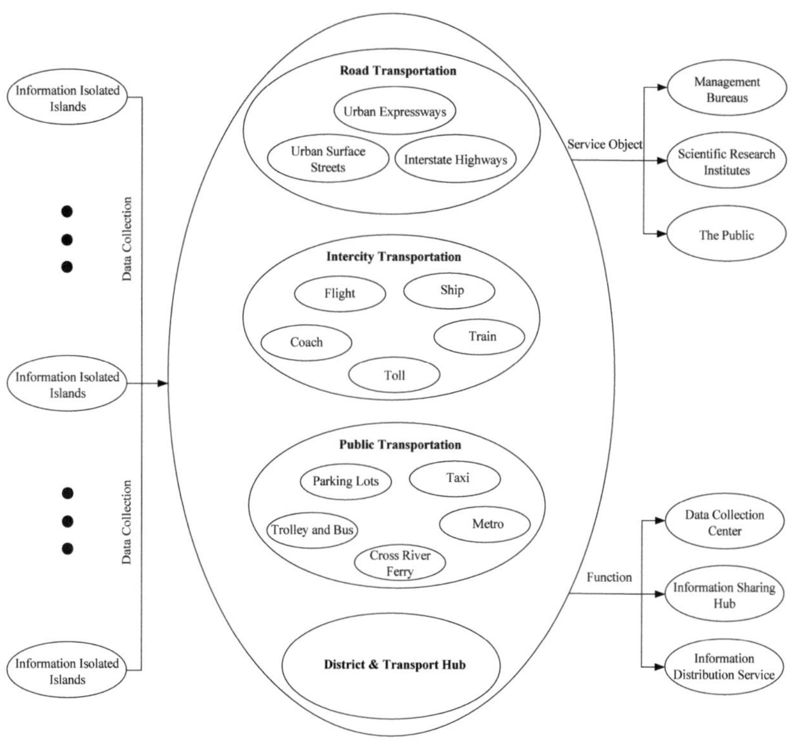

Fig. 2. The main structure of SITIC.

6. Conclusion

The chapter summarizes data driven approaches for traffic prediction in three parts. First, on the basis of classification of the main methods for traffic forecasting, the chapter aims to describe a large amount of literature of traffic forecasting models. And we focus on the decription of combined forecasts approaches that we believe represent the trend of the development of traffic forecasting in practice. Second, from the practical point of view, proper solutions to solve missing data problems are decribled, espertially the state space based approaches. Finally, from the perspective of dynamic traffic management, it presents the corresponding work and experience of traffic informatization in Shanghai.

7. References

Abdulhai, B.; Porwal, H.; & Recker, W. (1999). Short-term freeway traffic flow prediction using genetically optimized time-delay-based neural networks, *Proc. 78th TRB Annual Meeting*, Washington, D.C

Ahmed, M. S. & Cook, A. R. (1979). Analysis of freeway traffic time-series data by using Box-Jenkins techniques, *Transportation Research Record. 722*, Transportation Research Board, Washington, D.C., pp. 1-9

Bates, J. M. and Granger, C. W. J. (1969). The combination of forecasts, *Oper. Res. Quarterly.*, Vol.20, No.4, pp. 451-468

Brockwell, P. J. & Davis, R. A. (2002). *Introduction to Time Series and Forecasting*, Springer-Verlag, New York

Castillo,E.; Menéndez,J.M. & Cambronero,S.S. (2008). Predicting traffic flow using Bayesian networks, *Transportation Research Part B: Methodological.*, Vol.42, No.5, pp. 482-509

Castro-Neto M., Jeong Y. S., Jeong M. K., & Han L.D. (2009). Online-SVR for short-term traffic flow prediction under typical and atypical traffic conditions, *Expert Systems with Applications*, Vol. 36, No. 3, pp. 6164-6173

Chan, K.S.; Lam, W. H. K. & Xu, G. (2003). A Case Study on Short-term Travel Time Forecasting in Hong Kong, *Proceedings of the 8th Conference of Hong Kong Society for Transportation Studies*, Hong Kong, 13-14 December, pp. 444-451

Chan, K.S. & Lam, W. H. K. (2005). On-line Travel Time Forecasts for Real-time Traveller Information System in Hong Kong, *Proceedings of the 10th International Conference of Hong Kong Society for Transportation Studies*, 10 December, Hong Kong, pp. 94-102

Chang, G. L., and Su, C. C. (1995). "Predicting intersection queue with neural network models, *Transp. Res., Part C: Emerg. Technol.*, Vol.3, No.3, pp. 175-191

Chang, J.; Chowdhury, N. K. & Lee H. (2010). New travel time prediction algorithms for intelligent transportation systems, *Journal of Intelligent & Fuzzy Systems: Applications in Engineering and Technology*, Vol.21, No.1-2, pp. 5-7

Chen, H. B, Grant-Muller, S., Mussone, L., and Montgomery, F. (2001). A study of hybrid neural network approaches and the effects of missing data on traffic forecasting, *Neural Comput. & Applic.* Vol.10, No.3, pp. 277-286

Chen, D. W. & Zhang, J. P. (2005). Time series prediction based on ensemble ANFIS, *Proc. Int. Conf. Machine Learning and Cybernetics.*, Vol. 6, pp. 3552-3556

Chen, L. & Chen, C. L. P. (2007). Ensemble learning approach for freeway short-term traffic flow prediction, *Proc. IEEE Int. Conf. Sys. of Sys. Eng.*, pp. 1-6

Chen, S.Y.; Wang, W. & Zuylen, H.V. **(2010).** A comparison of outlier detection algorithms for ITS data, *Expert Systems with Applications*,Vol.37, No.2, pp. 1169-1178

Chien S. I. J. & Kuchipudi C. M., (2003). Dynamic travel time prediction with real-time and historic data, *J. Transp. Eng.*, Vol.129, No.6, pp. 608-616

Chrobok, R.; Kaumann, O.; Wahle, J. & Schreckenberg, M. (2004). Different methods of traffic forecast based on real data, *Eur. J. Oper. Res.* Vol.155, No.3, pp. 558-568

Clark, S. D.; Dougherty, M. S. & Kirby, H. R. (1993). The use of neural networks and time series models for short-term traffic forecasting: a comparative study, *Proc. 21st Summer Annual Meeting*, PTRC, Manchester, UK

Clark, S. (2003). Traffic prediction using multivariate nonparametric regression, *J. Transp. Eng.*, Vol.129, No.2, pp. 161-168

Clemen, R. T. (1989). Combining forecasts: a review and annotated bibliography, *Int. J. Forecast.*, Vol.5, pp. 559-583

Davis, G. A. & Nihan, N. L. (1991). Nonparametric regression and short-term freeway traffic forecasting, *J. Transp. Eng.*, Vol.117, No.2, pp. 178-188

De Menezes, L. M.; Bunn, D. W. & Taylor, J. W. (2000). Review of guidelines for the use of combined forecasts, *Eur. J. Oper. Res.*, Vol.120, pp. 90-204

Dempster, A. P.; Laird, N. M. & Rubin, D. B. (1977). Maximum-likelihood estimation from incomplete data via the EM algorithm, *J. Royal Statist. Soc., Series B.*, Vol. 39, pp. 1-38

Deng, S.; Hu J. M.; Wang Y., & Zhang, Y. (2009). Urban road network modeling and real-time prediction based on householder transformation and ajacent vector, *Lecture Notes in Computer Science*, Vol. 5553, pp. 899-908

Dharia, A. & Adeli, H. (2003). Neural network model for rapid forecasting of freeway link travel time, *Eng. Applic. Artif. Intell.*, Vol.16, No.7-8, pp. 607-613

Dia, H. (2001). An object-oriented neural network approach to short-term traffic forecasting, *Eur. J. of Oper. Res.* Vol.131, No.2, pp. 253-261

Dimitriou,L.; Tsekeris, T. & Stathopoulos,A., (2008). Adaptive hybrid fuzzy rule-based system approach for modeling and predicting urban traffic flow , *Transportation Research Part C: Emerging Technologies.*, Vol.16, No.5, pp. 554-573

Dickinson, J. P. (1973). Some statistical results in the combination of forecasts, *Oper. Res. Quarterly*, Vol.24, No.2, pp. 253-260

Ding, A.; Zhao X. & Jiao, L. (2002). Traffic flow time series prediction based on statistics learning theory." *Proc. IEEE 5th Int. Conf. Intell. Transp. Syst.*, pp. 727–730

Dong, S.; Li, R. M.; Sun, L. G., Chang, T. H.; & Lu, H. P. (2010). Short-term traffic forecast system of Beijing, *Journal of the Transportation Research Board*, Vol. 2193, pp. 116-123

Dougherty, M. S. & Cobbet, M. R. (1997). Short-term inter-urban traffic forecasts using neural networks, *Int. J. Forecast.*, Vol.13, pp. 21-31

Florio, L. & Mussone, L. (1996). Neural network models for classification and forecasting of freeway traffic flow stability, *Control Eng. Pract.*, Vol.4, No.2, pp. 153-164

Ghosh, B.; Basu, B. & O'Mahony, M. M. (2005). Time-series modelling for forecasting vehicular traffic flow in Dublin, *Proc. 85th TRB Annual Meeting*, Washington, D.C.

Ghosh, B.; Basu, B. & O'Mahony, M. M. (2007). Bayesian time-series model for short-term traffic flow forecasting, *J. Transp. Eng.*, Vol.133, No.3, pp. 180-189

Gilmore, J. E. & Abe, N. (1995). Neural network models for traffic control and congestion prediction, *IVHS J.*, Vol.2, No.3, pp. 231-252

Granger, C. W. J. & Ramanathan, R.(1984). Improved methods of forecasting, *J. Forecast.*, Vol. 3, pp. 197-204

Hamed, M. M., Al-Masaeid, H. R., and Said, Z. M. B. (1995). "Short-term prediction of traffic volume in urban arterials." *J. Transp. Eng.*, Vol.121, No.3, pp. 249-254

Hong, W. C. (2011). Traffic flow forecasting by seasonal SVR with chaotic simulated annealing algorithm, *Neurocomputing*, Vol. 74, No. 12-13, pp. 2096-2107

Hong, W.C.; Dong, Y.C.; Zheng, F.F. & Wei, S.Y. (2011). Hybrid evolutionary algorithms in a SVR traffic flow forecasting model , *Applied Mathematics and Computation*, Vol.217, No.15, pp. 6733-6747

Huang, H. (2007). Essays on combination of forecasts, Doctoral dissertation, Graduate Program in Economics, University of California, Riverside

Huang, S. & Sadek, A.W. (2009). A novel forecasting approach inspired by human memory: The example of short-term traffic volume forecasting, *Transportation Research Part C: Emerging Technologies*, Vol.17, No.5, pp. 510-525

Huisken, G. (2003). Soft-computing techniques applied to short-term traffic flow forecasting, *Systems Analysis Modelling Simulation*, Vol.43, No.2,pp. 165-173

Innamaa, S. (2000). Short-term prediction of traffic situation using MLP-Neural Networks, *Proc. 7th World Cong. Intell. Transp. Syst.*, Turin, Italy

Ishak, S. & Alecsandru, C. (2003). Optimizing traffic prediction performance of neural networks under various topological, input, and traffic condition settings, *Proc. 82nd TRB Annual Meeting*, Washington, D.C.

Jiang, X. M. & Adeli, H. (2004). Wavelet Packet-Autocorrelation Function Method for traffic flow pattern analysis, *Comput.-Aided Civ. & Inf. Eng.*, Vol.19, pp. 324-337

Jiang, X. M. & Adeli, H. (2005). Dynamic wavelet neural network model for traffic flow forecasting, *J. Transp. Eng.*, Vol.131, No.10, pp. 771-779

Jose, V. R. R. & Winkler, R. L. (2008). Simple robust averages of forecasts: some empirical results, *Int. J. Forecast.*, Vol.24, pp. 163-169

Kamarianakis, Y. & Prastakos, P. (2003). Forecasting traffic flow conditions in an urban network: Comparison of multivariate and univariate approaches, *Proc. 82nd TRB Annual Meeting*, Washington, D.C.

Kamarianakis, Y. & Prastakos, P. (2005). Space-time modeling of traffic flow, *Computers & Geosciences*, Vol.31, No.2, pp. 119-133

Khan, A. M. (2011), Bayesian predictive travel time methodology for advanced traveller information system, *Journal of Advanced Transportation*, 45: n/a. doi: 10.1002/atr.147

Kirby, H. R.; Watson, S. M.& Dougherty, M. S. (1997). Should we use neural network or statistical models for short-term motorway traffic forecasting? *Int. J. Forecast.*, Vol.13, No.1, pp. 43-50

Kwon, J.; Coifman, B. & Bickel, P. (2000). Day-to-Day Travel Time Trends and Travel Time Prediction from Loop Detector Data, *Transportation Research Record*. 1717, *TRB*, pp. 120-129

Kwon, J. & Petty, K. (2005). A Travel Time Prediction Algorithm Scalable to Freeway Networks with Many Nodes with Arbitrary Travel Routes, *Transportation Research Record*. 1935, *TRB*, pp. 147-153

Lam, W. H. K. & Xu, J. (2000). Estimation of AADT from short period counts in Hong Kong—A comparison between neural network method and regression analysis, *J. Adv. Transp.*, Vol. 34, No. 2, pp. 249–268

Lam, W. H. K.; Chan K.S. & Shi, J.W.Z. (2002). A Traffic Flow Simulator for Short-term Travel Time Forecasting, *Journal of Advanced Transportation*, Vol. 36, No. 3, pp. 231-242

Lam, W. H. K. & Chan, K.S. (2004). Short-term forecasting of travel time and reliability, *Proceedings of the International Workshop on Behavior in Networks*, ed. Seungjae Lee, William H.K. Lam and Yasuo Asa, July 22-23, The University of Seoul, Seoul, Korea, pp. 127-135

Lam, W. H. K.; Chan, K.S.; Tam, M.L. & Shi, J. W. Z. (2005). Short-term Travel Time Forecasts for Transport Information System in Hong Kong, *Journal of Advanced Transportation*, Vol. 39, No. 3, pp. 289-305

Lam, W. H. K.; Tang, Y. F.; Chan, K. S. & Tam, M. L. (2006). Short-term hourly traffic forecasts using Hong Kong annual traffic census, *Transp.*, Vol. 33, No. 3, pp. 291-310

Lam, W. H. K.; Tang, Y. F. & Tam, M. L. (2006). Comparison of two non-parametric models for daily traffic forecasting in Hong Kong, *Int. J. Forecast.*, Vol. 25, No. 3, pp. 173-192

Lam, W. H. K. (2008). Short-term Forecasting of Travel Times for Hong Kong Incident Management, *Proceedings of the ITS Hong Kong Forum 2008 – Incident Management System (IMS) Technologies and Applications*, 24 September, Hong Kong, pp. 35-50

Lam, W. H. K.; Tam, M.L.; Sumalee, A.; Li, C.L.; Chen, W.; Kwok, S.K.; Li, Z.L. & E.W.T. (2008). Ngai Incident Detection based on Short-term Travel Time Forecasting, *Proceedings of the 13th International Conference of Hong Kong Society for Transportation Studies*, 13-15 December, Hong Kong, pp. 83-92

Ledoux, C. (1997). An urban traffic flow model integrating neural networks, *Transp. Res., Part C: Emerg. Technol.*, Vol. 5, No. 5, pp. 287-300

Lee, S. & Fambro, D. B. (1999). Application of subset autoregressive integrated moving average model for short-term freeway traffic volume forecasting, *Transportation Research Record*. 1678, *TRB*, pp. 179-188

Lee, W. H., Tseng, S. S. & Tsai, S. H. (2009). A knowledge based real-time travel time prediction system for urban network, Expert Systems with Applications, Vol. 36, No. 3, Part 1, pp. 4239-4247

Lee, Y. (2009). Freeway travel time forecast using artificial neural networks with cluster method, *Proc. of the 12th International Conference on Information Fusion*, Seattle, pp: 1331-1338

Levin, M. & Tsao, Y. D. (1980). On forecasting freeway occupancies and volumes,*Transportation Research Record*. 773, *TRB*, Washington, D.C., pp. 47-49

Lingras, P.; Sharma, S.C. & Osborne, P. (2000). Traffic volume time-series analysis according to the type of road use, *Comput.-Aided Civ. & Inf. Eng.*, Vol. 15, No. 5, pp. 365-373

Lingras, P. & Mountford, P. (2001). Time delay neural networks designed using genetic algorithms for short-term inter-city traffic forecasting, LNAI 2070, IEA/AIE, Vienna, pp.290-299

Lu,Y.; Hu, J.M.; Xu, J. & Wang, S.N. (2009). Urban Traffic Flow Forecasting Based on Adaptive Hinging Hyperplanes, *Proceedings of the International Conference on Artificial Intelligence and Computational Intelligence*, pp. 658-667

Lyons, G. D.; McDonald, M.; Hounsell, N. B.; Williams, B.; Cheese, J. & Radia, B. (1996). Urban traffic management: the viability of short-term congestion forecasting using artificial neural networks, *Proc. 24th European Transport Forum*, PTRC, pp.1-12

Min W., & Wynter L. (2011). Road traffic prediction with spatial-temporal correlations, *Transportation Research Part C: Emerging Technologies,* Vol. 19, No. 4, pp. 606-616

Min X. Y.; Hu, J. M.; & Zhang, Z. (2010). Urban traffic network modeling and short-term traffic flow forecasting based on GSTARIMA model, *The 13th International IEEE Conference on Intelligent Transportation Systems,* pp: 1535-1540

Nath,R.P.D.; Lee,H.J.; Chowdhury, N.K. & Chang J.W. (2010). Modified K-means clustering for travel time prediction based on historical traffic data, *Proceedings of the 14th international conference on Knowledge-based and intelligent information and engineering systems,* PP. 511-521

Neto,M.C.; Jeong,Y.S.; Jeong,M.K. & Han, L.D. (2009). Online-SVR for short-term traffic flow prediction under typical and atypical traffic conditions, *Expert Systems with Applications,* Vol. 36, No. 3, pp. 6164-6173

Ni, D. H.; Leonard II, J. D.; Guin, A. & Feng, C. X. (2005). Multiple imputation scheme for overcoming the missing values and variability issues in ITS data, ASCE *J. Transp. Eng.,* Vol. 131, No. 12, pp. 931-938

Nihan, N. L. & Holmesland, K. O. (1980). Use of the Box and Jenkins time series technique in traffic forecasting, *Transportation,* Vol. 9, No. 2, pp. 125-143

Okutani, I. & Stephanedes, Y. J. (1984). Dynamic prediction of traffic volume through Kalman filtering theory, *Transp. Res., Part B: Methodol.,* Vol. 18, No. 1, pp. 1-11

Park, B. (2002). Hybrid neuro-fuzzy application in short-term freeway traffic volume forecasting, *Transportation Research Record.* 1802, *TRB,* Washington, D.C., pp. 190-196

Park, B.; Messer, C. J. & Urbanik, T., II. (1998). Short-term freeway traffic volume forecasting using radial basis function neural network, *Transportation Research Record.* 1651, *TRB,* Washington, D.C., pp. 39-47

Park, D. & Rilett, L. R. (1998). Forecasting multiple-period freeway link travel times using modular neural networks, *Transportation Research Record.* 1617, *TRB,* pp. 63-70

Park, D.; Rilett, L. R. & Han, G. (1999). Spectral basis neural networks for real-time travel time forecasting, *J. Transp. Eng.,* Vol. 125, No. 6, pp. 515-523

Qiao, F.; Yang, H. & Lam, W. H. K. (2001). Intelligent Simulation and Prediction of Traffic Flow Dispersion, *Transportation Research-A,* Vol. 35, No. 9, pp. 843-863

Quek, C.; Pasquier, M. & Lim, B. B. S. (2006). POP-TRAFFIC: A novel fuzzy neural, approach to road traffic analysis and prediction, *IEEE Trans. Intell. Transp. Syst.,* Vol. 7, No. 2, pp. 133-146

Quek, C.; Pasquier, M. & Lim, B. (2009). A novel self-organizing fuzzy rule-based system for modelling traffic flow behaviour, *Expert Systems with Applications.,* Vol. 36, No. 10, pp. 12167-12178

Rice, J. & Van Zwet, E. (2004). A simple and effective method for predicting travel times on freeways, *IEEE Trans. Intell. Transp. Syst.,* Vol.5, No.3, pp. 200-207

Ross, P. (1982). Exponential filtering of traffic data, *Transportation Research Record.* 869, *TRB,* Washington, D.C., pp. 43-49

Schadschneider, A.; Wolfgang Knospe, W.; Santen, L. & Schreckenberg,M. (2005). Optimization of highway networks and traffic forecasting, *Physica A: Statistical Mechanics and its Applications,* Vol.346, No.1-2, pp. 165-173

Schafer, J. L. (1997). *Analysis of Incomplete Multivariate Data.* New York: Chapman & Hall

Smith, B. L. (1995). Forecasting freeway traffic flow for intelligent transportation system applications, *Doctoral dissertation.* Department of Civil Engineering, University of Virginia, Charlottesville

Smith, B. L. & Demetsky, M. J. (1994). Short-term traffic flow prediction: Neural network approach, *Transportation Research Record*. 1453, *TRB*, Washington, D.C., pp. 98-104

Smith, B. L. & Demetsky, M. J. (1997). Traffic flow forecasting: Comparison of modeling approaches, *J. Transp. Eng.*, Vol.123, No.4, pp. 261-266

Smith, B. L.; Williams, B. M. & Oswald, R. K., (2000). Parametric and nonparametric traffic volume forecasting, *Proc. Transportation Research Board Annual Meeting, TRB*, Washington, D.C., Reprint 00-817

Smith, B. L.; Williams, B. M. & Oswald, R. K. (2002). Comparison of parametric and nonparametric models for traffic flow forecasting, *Transp. Res., Part C: Emerg. Technol.*, Vol.10, No.4, pp. 03-321

Stathopoulos, A. & Karlaftis, M. G. (2001a). Temporal and spatial variations of real-time traffic data in urban areas, *Transportation Research Record*. 1768, *TRB*, pp. 135-140

Stathopoulos, A. & Karlaftis, M. G. (2001b). Spectral and cross-spectral analysis of urban traffic flows, *Proc. 4th IEEE Conf. Trans. Intell. Transp. Syst.*, Oakland, CA, pp. 820-825

Stathopoulos, A. & Karlaftis, M. G. (2003). A multivariate state-space approach for urban traffic flow modeling and prediction, *Transp. Res., Part C: Emerg. Technol.*, Vol.11, No.2, pp. 21-135

Sun, H.; Liu, H. X.; Xiao, H.; He, R. R. & Ran, B. (2003). Short-term traffic forecasting using the local linear regression model, *Proc. 82nd TRB Annual Meeting*, Washington, D.C

Sun, S. L & Zhang, C. S. (2007). The selective random subspace predictor for traffic flow forecasting, *IEEE Trans. Intell. Transp. Syst.*, Vol.8, No.2, pp. 367-373

Sun, S. L.; Zhang, C. S. & Yu, G. Q. (2006). A Bayesian network approach to traffic flow forecasting, *IEEE Trans. Intell. Transp. Syst.*, Vol.7, No.1, pp. 124-132

Sun, S. L.; Zhang, C. S. & Zhang, Y. (2005). Traffic flow forecasting using a spatio-temporal Bayesian network predictor, *Lect. Notes Comput. Sci.*, Vol.3697, pp. 273-278

Tan, M.C.; Wong, S.C.; Xu, J.M.; Guan, Z.R. & Zhang, P. (2009). An aggregation approach to short-term traffic flow prediction, *IEEE Transactions on Intelligent Transportation Systems*,Vol. 10, No. 1, pp 60-69

Tang, Y.F. & Lam, W. H. K. (2001). Validation of Short-term Prediction for Annual Average Daily Traffic in Hong Kong, *Proceedings of the 6th Conference of Hong Kong Society for Transportation Studies*, Sheraton Hotel, Hong Kong, 1 December, pp. 141-151.

Tang, Y.F. Lam, W. H. K. & Ng, P. L. P. (2003). Comparison of Four Modeling Techniques for Short-term AADT Forecasting in Hong Kong, *ASCE Journal of Transportation Engineering*, Vol. 129, No. 3, pp 271-277

Tam, M.L. & Lam, W. H. K. (2009). Short-Term Travel Time Prediction for Congested Urban Road Networks, *Proceedings of the 88th Transportation Research Board Annual Meeting*, 11-15 January, Washington D.C., U.S.A., Paper no. 09-2313

Tanner, M. A. & Wong, W. H. (1987). The calculation of posterior distributions by data augmentation, *J. Amer.* Stat. *Assoc.*, Vol. 82, pp. 528–550

Tebaldi, C.; West, M. & Karr, A. F. (2002). Statistical analysis of freeway traffic flows, *Int. J. Forecast.*, Vol.21, No.1, pp. 39-68

Timmermann, A. (2006). Forecast combinations, Handbook of Economic Forecast., Amsterdam: Elsevier, pp. 135-196

Thomas, T.; Weijermars, W. & Berkum, E.V. (2010). Predictions of urban volumes in single time series, *IEEE Transactions on Intelligent Transportation Systems*, Vol.11, No.1, pp. 71-80

Tsekeris, T., & Stathopoulos, A. (2010). Short-term prediction of urban traffic variability: stochastic volatility modeling approach, *ASCE Journal of Transportation Engineering*, Vol. 136, No. 7, pp. 606-613

Turochy, R. E. (2006) Enhancing short-term traffic forecasting with traffic condition information, *J. Transp. Eng.*, Vol.132, No.6, pp. 469-474

Van Arem, B.; Kirby, H. R.; Van Der Vlist, M. J. M. & Whittaker, J. C. (1997). "Recent advances and applications in the field of short-term traffic forecasting." *Int. J. Forecast.*, Vol.13, No.1, pp. 1-12

Vanajakshi, L. & Rilett, L. R. (2004). A comparison of the performance of artificial neural network and support vector machines for the prediction of traffic speed, *IEEE Intell. Vehicles Symp.*, Parma, Italy, pp. 194-199

Van Lint, J. W. C.; Hoogendoorn, S. P. & Van Zuylen, H. J. (2002). Freeway travel time prediction with state-space neural networks: modeling state-space dynamics with recurrent neural networks, *Proc. 81st TRB Annual Meeting*, Washington, D.C.

Van Lint J. W. C.; Hoogendoorn S. P. & Van Zuylen H. J. (2005). Accurate freeway travel time prediction with state-space neural networks under missing data, *Transp. Res., Part C: Emerg. Technol.*, Vol.13, No.5-6, pp. 47-369

Vlahogianni, E. I.; Golias, J. C. & Karlaftis, M. G. (2004). Short-term forecasting: Overview of objectives and methods, *Transport Rev.*, Vol.24, No.5, pp. 533-557

Vlahogianni, E. I.; Karlaftis, M. G. & Golias, J. C. (2005). Optimized and meta-optimized neural networks for short-term traffic flow prediction: A genetic approach, *Transp. Res., Part C: Emerg. Technol.*, Vol.13, No.2, pp. 211-234

Vythoulkas, P. C. (1993). Alternative approaches to short-term traffic forecasting for use in driver information systems, *Proc. 12th Int. Symp. Traffic Flow Theory and Transportation*, C. F. Daganzo, ed., Berkeley, CA.

Wei, C. H.; & Lee, Y. (2007). Development of freeway travel time forecasting models by integrating different sources of traffic data, *IEEE Trans. on Vehicular Tech.*, Vol. 56, No. 6, pp. 3682-3694

Wei, L. Y., Fang Z. W., & Luan, S. (2009). Travel time prediction method for urban expressway link based on artificial neural network. *The 5th International Conference on Natural Computation*, pp: 358-362

Whittaker, J.; Garside, S. & Lindveld, K. (1997). Tracking and predicting a network traffic process, *Int. J. Forecast.*, Vol.13, No.1, pp. 51-61

Wild, D. (1997). Short-term forecasting based on a transformation and classification of traffic volume time series, *Int. J. Forecast.*, Vol.13, No.1, pp. 63-72

Williams, B. M. (1999). Modeling and forecasting vehicular traffic flow as a seasonal stochastic time series process, *Doctoral dissertation*. Department of Civil Engineering, University of Virginia, Charlottesville

Williams, B. M. (2001). Multivariate vehicular traffic flow prediction: an evaluation of ARIMAX modeling, *Proc. 80th TRB Annual Meeting, Mira Digital Publishing*, Washington D.C.

Williams, B. M.; Durvasula, P. K. & Brown, D. E. (1998). Urban freeway traffic flow prediction: Application of seasonal autoregressive integrated moving average and exponential smoothing models, *Transportation Research Record*. 1644, *TRB*, Washington, D.C., pp. 132-141

Williams, B. M. & Hoel, L. A. (2003). Modeling and forecasting vehicular traffic flow as a seasonal ARIMA process: Theoretical basis and empirical results, *J. Transp. Eng.*, Vol.129, No.6, pp. 664-672

Wu, C. H.; Ho, J. M. & Lee, D. T. (2004). Travel-time prediction with support vector regression, *IEEE Trans. Intell. Transp. Syst.*, Vol.5, No.4, pp. 276-281

Yang, M.L.; Liu, Y.G. & You, Z.S. (2010). The reliability of travel time forecasting, *IEEE Transactions on Intelligent Transportation Systems*, Vol.11, No.1, pp. 162-171

Yang, Y. (2004). Combining forecasting procedures: some theoretical results, *Econometric Theory.*, Vol.20, pp. 176-222

Yasdi, R. (1999). Prediction of road traffic using a neural network approach, *Neural Comput. Appl.*, Vol.8, No.2, pp. 135-142

Yin, H.; Wong, S. C.; Xu, J. & Wong, C. K. (2002). Urban traffic flow prediction using a fuzzy-neural approach, *Transp. Res., Part C: Emerg. Technol.*, Vol.10, No.2, pp. 85-98

You, J. & Kim, T. J. (2000). Development and evaluation of a hybrid travel time forecasting model, *Transp. Res., Part C: Emerg. Technol.*, Vol.8, No.1-6, pp. 231-256

Yu, L.; Wang, S. & Lai, K. K. (2005). A novel nonlinear ensemble forecasting model incorporating GLAR and ANN for foreign exchange rates, *Comp. & Oper. Res.*, Vol.32, No.10, pp. 2523-2541

Yu, L.; Wang, S.; Lai, K. K. & Nakamori, Y. (2005). Time series forecasting with multiple candidate models: selecting or combining? *J. Sys. Sci. & Complexity.*, Vol.18, No.1, pp. 1-18

Yun, S. Y.; Namkoong, S.; Rho, J. H.; Shin, S. W. & Choi, J. U. (1998). A performance evaluation of Neural Network Models in Traffic Volume Forecasting, *Math. Comput. Modell.*, Vol.27, No.9-11, pp. 293-310

Zhang, Y. & Liu, Y.C. (2008). A Novel Approach to Forecast Weakly Regular Traffic Status, *11th International IEEE Conference on Intelligent Transportation Systems (ITSC 2008)*, pp: 998-1002, Beijing, China

Zhang, Y. & Liu, Y.C. (2009a). Data Imputation using Least Squares Support Vector Machines in Urban Arterial Streets, *IEEE Signal Processing Letters*, Vol. 16, No. 5, pp: 414-417

Zhang, Y. & Liu, Y.C. (2009b). Missing Traffic Flow Data Prediction using Least Square Support Vector Machines in Urban Arterial Streets, *IEEE 2009 Symposium on Computational Intelligence and Data Mining (SSCI2009)*, Sheraton Music City Hotel, Nashville, TN, USA, 30 Mar. - 2 Apr.

Zhang, Y. & Liu, Y.C. (2009c). Traffic Forecasting using Least Squares Support Vector Machines, *Transportmetrica*, Vol. 5, No. 3, pp: 193-213

Zhang, Y. & Liu, Y.C. (2009d). Traffic Forecasts using Interacting Multiple Model Algorithm, *Lecture Notes in Artificial Intelligence (LNAI)*, Vol. 5579, pp: 360-368

Zhang, Y. & Liu, Y.C. (2009e). Application of Combined Forecasting Models to Intelligent Transportation Systems, *Opportunities and Challenges for Next Generation Applied Intelligence, 22nd International Conference on Industrial, Engineering & Other Applications of Applied Intelligent Systems (IEA-AIE 2009)*, Vol. 214, pp: 181-186

Zhang, Y. & Liu, Y.C. (2009f). Comparison of Parametric and Nonparametric Techniques for Non-peak Traffic Forecasting, *Proceedings of World Academy of Science, Engineering and Technology, Vol. 39, International Conference on Computational Statistics and Data Analysis (ICCSDA 2009)*, pp: 242-248

Zhang, Y. & Liu, Y.C. (2010). Analysis of Peak and Non-peak Traffic Forecasts using Combined Models, *Journal of Advanced Transportation*, Vol. 17, pp: 1-17

Zhang, Y.; Shi, W.H. & Liu, Y.C. (2010). Comparison of Several Traffic Forecasting Methods based on Travel Time Index Data on Weekends, *Journal of Shanghai Jiao Tong University(Science)*, Vol. 12, No. 2, pp: 188-193

Zhang, C.; Sun, S. & Yu, G. (2004). A Bayesian network approach to time series forecasting of short-term traffic flows, *Proc. 7th IEEE Int. Conf. Intell. Transp. Syst. (ITSC2004)*, Washington, D.C., pp. 216-221

Zhang, H.; Ritchie, S. G. & Lo, Z. P. (2000). Macroscopic modeling of freeway traffic using an artificial neural network, *Transportation Research Record.*, 1588, *TRB*, pp. 110-119

Zhang, X. Y., & Rice, J. A. (2003). "Short-term travel time prediction." *Transp. Res., Part C: Emerg. Technol.*, 11(3-4), 187-210

Zheng, W. Z.; Lee, D. H. & Shi, Q. X. (2006). Short-term freeway traffic flow prediction: Bayesian combined neural network approach, *J. Transp. Eng.*, Vol. 132, No.2, pp. 114-121

Zhong, M.; Sharma, S. C. & Lingras, P.(2004). Genetically designed models for accurate imputations of missing traffic counts, *Transp. Res. Rec.* 1879, *J. Transp. Res. Board, TRB*, Washington, D. C., pp. 71-79